QC
178
F 8.6

Fundamental Principles of
General Relativity
Theories

LOCAL AND GLOBAL ASPECTS OF GRAVITATION
AND COSMOLOGY

Fundamental Principles of General Relativity Theories

LOCAL AND GLOBAL ASPECTS OF GRAVITATION AND COSMOLOGY

Hans-Jürgen Treder
Horst-Heino von Borzeszkowski
Zentralinstitut für Astrophysik
Potsdam-Babelsberg, GDR

Alwyn van der Merwe
Wolfgang Yourgrau
University of Denver
Denver, Colorado, USA

Plenum Press · New York and London

Library of Congress Cataloging in Publication Data

Main entry under title:

Fundamental principles of general relativity theories.

Includes index.
1. General relativity (Physics) I. Treder, Hans-Jürgen.
QC173.6.F86   530.1'1   80-14826
ISBN 0-306-40405-2

© 1980 Plenum Press, New York
A Division of Plenum Publishing Corporation
227 West 17th Street, New York, N.Y. 10011

Printed in the United States of America

Wolfgang Yourgrau 1908–1979

The completion of this book, which would normally have been cause for celebration, became instead a sorrowful occasion because of the untimely passing away of our coauthor Wolfgang Yourgrau on July 18. By way of paying tribute to this remarkable individual, we should like to recall very briefly some of the highlights of his extraordinary career.

Wolfgang Yourgrau was born on November 16, 1908 into a family of Belgian and German ancestry. He attended the Werner-Siemens Realgymnasium in Berlin and then went on to study theoretical physics, mathematics, and biology at the local von Humboldt University, at a time when celebrated physicists such as von Laue, Einstein, and Schrödinger graced its faculty. Serving first as a tutor in natural philosophy and then as an assistant to Schrödinger, he earned his Dr. phil. *magna cum laude* in 1932.

Compelled by the Nazis to leave Germany the next year, he settled, after a period of traveling and lecturing, in Palestine, where he briefly published and edited the independent weekly *Orient* and found himself again in the thick of political controversy.

Wolfgang Yourgrau 1908–1979

Some time after the United States entered the War, Yourgrau joined the United States Office of Strategic Services, which carried out intelligence operations behind enemy lines. At War's end he became head of the Department of Logic and Scientific Method at the School of Higher Studies in Jerusalem and, subsequently, Acting Dean of its Faculty of Arts and Sciences.

From Israel he emigrated to South Africa, where for a decade he taught at the Universities of Cape Town, Witwatersrand, and the Province of Natal. In 1959 he moved to the United States, first to accept the position of Research Professor at the Minnesota Center for the Philosophy of Science, and then to become Chairman of the Department of History of Science at Smith College in Northampton, Massachusetts. In 1963 he accepted a permanent position as professor of History and Philosophy of Science at the University of Denver.

The range of Yourgrau's publications was prodigious and made him known to an exceptionally large and diverse group of scholars. Although most of his papers are devoted to problems in physics, a large fraction deals with philosophic issues, while some others treat matters of a biographical, literary, or political nature. Of the many books he coauthored or coedited, *Variational Principles in Dynamics and Quantum Theory* and *A Treatise on Irreversible and Statistical Thermophysics* are perhaps the best known. In 1969 he founded, with Henry Margenau, the international journal *Foundations of Physics*, which he coedited until his death. He was a recipient of numerous distinctions and honors, which included the Einstein Medal awarded to him in 1970.

Aside from his passion for life and his profound humanism, Wolfgang Yourgrau will be remembered by us for his positive thinking and the intellectual excitement which he so superbly managed to convey to his coworkers. It is to this driving force

6

that also the present monograph owes both its conception and its final appearance in print. We, who were privileged to be among his close friends, therefore deem it most appropriate to dedicate this book to his memory.

H.-J. Treder

H.-H. von Borzeszkowski

A. van der Merwe

Preface

The present monograph is not a self-contained introductory text. Instead it presupposes to a large extent that the reader is not only thoroughly familiar with the *special* theory of relativity, but that he or she has studied the standard aspects of the *general* theory, as well.

Starting from local and global formulations of the principles of inertia and relativity, we discuss the microscopic and telescopic aspects of gravitation. Our central goal has been to demonstrate that the foundations of gravitational theory laid by Newton and Einstein imply questions on the relation between the micro- and macrocosm. The discussions surrounding these physical points can be rather well understood without an elaborate mathematical formalism.

All the same, we have attempted to make the main theme of our presentation accessible also to readers outside the circle of pundits by including two appendixes of a largely instructional nature. Appendix A gives a brief review of the basic concepts of four-dimensional spaces, for the convenience of readers who need such a recapitulation, while Appendix B deals with the more exotic notions of tetrad theory, which admittedly stands in wider need of elucidation. Both appendixes are meant in any event to serve the useful purpose of establishing our notation and collecting formulas for easy reference in the main body of the book. The general reader may accordingly find it helpful first to peruse one or both of the appendixes before turning to the Introduction and Chapter 1.

H.-J. Treder
H.-H. von Borzeszkowski
A. van der Merwe
W. Yourgrau

Contents

Contents

Introduction

The general-relativistic theory of gravitation is based on two major principles: One expresses the characteristic property of universal gravitation, while the other provides the key to the incorporation of the theory of gravitation into relativistic physics. Both of them can be enunciated in different versions with varying degrees of stringency, and the various formulations of the general-relativistic theory of gravitation are distinguishable from one another particularly by the manner in which they individually seek the implementation of these principles.

Einstein set forth both fundamental principles in their full generality at the time he laid the foundations of his general theory of relativity.

The first principle, *Einstein's principle of equivalence* (Einstein, 1907), postulates the equivalence of inertia and gravitation. This principle, which experiments dating from Galileo to Eötvös and Dicke have substantiated with ever-increasing accuracy as the law of the universal proportionality of inertial mass and passive gravitational mass, through application of Newton's third law becomes the *strong equivalence principle,* asserting the local indistinguishability of inertia and gravitation. The equivalence principle—in contrast to all field theories of the special theory of relativity—implies that the "gravitational charge" M is neither a universal nor a constant quantity. Such a universal constancy belongs only to the "specific gravitational charge," the ratio M/m of gravitational mass to inertial mass.

As Einstein recognized, this peculiarity of gravitation precludes the incorporation of Newton's theory of gravitation into the system of special-relativistic field theories and thus requires the broadening of the special relativity principle into a *general relativity principle.* Following Einstein, this principle has been understood as asserting, on the one hand, the *covariance* of all laws of physics under space–time coordinate transformations of the *Einstein group* and, on the other, the *local Lorentz invariance* of physical laws, that is, their independence from the choice of local systems of reference. (See Appendix B for the distinction between coordinate systems and systems of reference.) The principle of local Lorentz invariance has an obvious advantage over its dual, the covariance principle, in that it is physically more transparent.

As Einstein first showed, the general relativity principle also makes possible the constructive mathematical formulation of the equivalence principle. The different versions of the equivalence principle can in fact be understood with reference to the different possibilities for stating the general relativity principle.

Einstein's general theory of relativity of 1915 furnishes the maximum fulfillment of the equivalence and general relativity principles in their sharpest formulations. Taken together, these principles require a *geometrization of the gravitational field.* The flat special-relativistic space–time of Minkowski is thus converted into a curved Riemannian manifold. Einstein's general theory of relativity then asserts, in its strongest expression of the equivalence and general relativity principles, that gravitation is given exclusively by the Riemannian metric g_{ik} of the space–time world, and that the Lorentz-invariant and Einstein-covariant formulation of physical equations in terms of this metric already takes full account of the influence that gravitation has on all physical phenomena.

The necessary requirement of the correspondencewise transition of this geometrical gravitational theory into Newtonian gravitodynamics uniquely restricts the field equations for the gravitational field g_{ik} itself to those of Einstein:

$$R_{ik} - \tfrac{1}{2}g_{ik}R = -\varkappa T_{ik}$$

with

$$\varkappa = \frac{8\pi f}{c^4}$$

if f and $\varkappa$ denote the gravitational constants of Newton and Einstein, respectively. Among all mathematically possible general-relativistic gravitational theories, the Einsteinian gravitational theory is thus the one that comes closest to the theory of Newton. Attempts to develop modified or more general theories of gravitation, transcending that of Einstein, consequently involve a further departure from the basic ideas of Newton's theory and are necessarily based on a weakening or splitting of both Einstein principles.

A renunciation of the strong equivalence in favor of the *Galilei–Eötvös* principle, which demands the mere equality of inertial and passive gravitational masses, makes it possible to introduce in gravitational theory more field quantities than the ten components g_{ik} of Riemann's space-time metric for the description of the gravitational field. In this manner, the *bimetric*, the *scalar-tensor* and, above all, also the *reference-tetrad* theories of gravitation are obtained. The reference-tetrad theories of gravitation represent a wide-ranging extension of Einstein's general theory of relativity when all 16 components $h^A{}_i$ of the reference tetrad are employed to serve as potentials of the gravitational field. In the gravitational theories with *Cartan–Einstein distant parallelism*, which results from this approach, one can still represent the effect of gravitation on matter in a locally Lorentz-covariant manner. But the same does not hold any more for the free gravitational field itself, which now violates general relativity in the sense that it is not invariant with respect to the local Lorentz group.

All the gravitational field theories that are here considered contain two universal constants, the gravitational constant f of Newton and c^2, the square of the special-relativistic light velocity. They are the constants that Newtonian theories of gravitation ought maximally to contain. By confining ourselves to these constants and by invoking Einstein's principles, the structure of the gravitational equations becomes uniquely determined in their linear approximation. Indeed, the linear gravitational equations must be identical with the corresponding equations of Einstein, if on the right-hand side of the field equations one introduces, as sources of the gravitational field, Hilbert's special-relativistic metric *matter tensor* T_{ik}.

A far-reaching generalization of these gravitational theories results if one introduces, besides f and c, another universal constant, representing a microscopic length l or a microscopic area l^2, into the theory. In this approach it is required that Einstein's gravitational theory (or its generalizations here under consideration) be recovered in the limiting case $l \to 0$. With the help of the new constant l^2, it becomes possible to supplement Einstein's second-order gravitational equations by terms containing higher-order derivatives. The *Einstein vacua*, i.e., the Einstein spaces $R_{ik} = \lambda g_{ik}$, then appear as special solutions of the field equations outside matter. It is thus also possible to modify the general theory of relativity without altering the Einstein vacuum. This means that, on the right-hand side of Einstein's equations, Hilbert's matter tensor T_{ik} can be augmented by further tensors, which vanish when l^2 or f tends to zero.

The general-relativistic gravitational field equations with higher-order derivatives, as well as the Einstein equations with a generalized matter tensor, basically violate Einstein's weak equivalence principle: The general-relativistic representation of the matter field equations comprises here not only the influence of the Einstein–Newton gravitation on matter, but also the influence of additional universal interactions of a microscopic nature, which vanishes as l tends to zero, a so-called *supergravitation*.

Einstein's general theory of relativity and its physical field modifications represent attempts to extend *Faraday's principle* of the local nature of all interactions to Newtonian gravitation. Indeed, Einstein's geometrical interpretation of the equivalence principle realizes the locality postulate, in so far as it locally reduces gravitation to inertia and identifies both with the local world metric g_{ik}.

Both the equations of motion of Newtonian mechanics and the field equations of Einstein are differential equations. Naturally the differential equations do not fully determine either motions or fields; to that end, we furthermore need "integration constants," which the theory itself does not provide. These constants are in the theory of Einstein, in particular, associated with the boundary conditions at infinity or, more generally, the topological *global structure* of Riemannian space–time. This global structure takes the place of Newton's absolute space, as that reference frame with respect to which dynamics is to be formulated.

Introduction

Actually the determination of the global frame of reference is, with Newton as well as with Einstein, more essential than the dynamic equations themselves. According to Newton's first law of motion, absolute space (or, more generally, Galilean inertial systems) provides us with the simplest description of force-free inertial motion; analogously, in the theory of Einstein, the asymptotic Minkowski space furnishes the zero-order approximation to the solution of Einstein's gravitational equations. The local gravitational fields determine only in higher-order approximation the metric in the framework of a primarily given global structure.

The critics of Newtonian mechanics, from Leibniz and Huygens to Hertz, Mach, Poincaré, and Einstein himself, have, however, pointed out that Newton's celestial mechanics was in fact formulated with reference to a *fundamental astronomical system*, the "fixed stars" or the cosmos. Einstein's principle of general relativity also contained the germ of *Mach's postulate* (1883) on the "relativity of inertia." Indeed, it is true that the equivalence between inertia and gravitation, which in a local theory of gravitation is interpreted as reducing gravitation to inertia (that is, eliminating Newton's second law in favor of his first), can also be interpreted, conversely, as reducing inertia to gravitation (and thus eliminating Newton's first law against his second).

This second interpretation, the *Mach–Einstein doctrine*, postulates that all inertial forces acting on a given object depend on the gravitational interactions existing between the object and distant cosmic masses. The *inertia-free gravitodynamics*, resulting from this *telescopic* (after Planck) action, then presents itself as an action-at-a-distance theory in an extreme form. The corresponding "general relativity principle," the *Mach–Einstein principle of the relativity of inertia*, accordingly generalizes not the global Lorentz invariance but the global Galilei invariance.

Introduction

Inertia-free gravitodynamics leads us mathematically from the three-dimensional space of classical physics into the $3N$-dimensional Lagrange–Hertz configuration space of N cosmic particles (point masses). In this configuration space, the Mach–Einstein doctrine then appears as a system of nonholonomic Hertzian conditions connecting the particle momenta (Hertz, 1894). These conditions are such that they require and make possible the elimination from dynamics of all inertial terms, while simultaneously prohibiting the Hertzian configuration space from resolving into the product of the three-dimensional single-particle spaces.

It should furthermore be noted that inertia-free gravitodynamics, since it renounces Newton's first law, is non-Newtonian in the extreme. Mach's principle, for the determination of physical reference frames through cosmic masses, provides, however, the possibility for specifying cosmic frames of reference in which the celestial mechanics of one- and two-body problems can be formulated, after Newton and Einstein, in a three-dimensional space whose Riemannian metric depends only on the local gravitational field.

Mach's principle establishes in fact the connection between gravitodynamics and *post-Newtonian* celestial mechanics. Thus, let the general-relativistic problem of motion be described, in the sense of Einstein, Infeld, and Hoffmann or Fock, with reference to the fundamental astronomical system, and let this reference frame be realized, in the sense of Mach's principle, by the relativistic cosmos. Comparison of the Mach–Einstein relativity of inertia with general relativity then determines a decisive cosmological quantity, the *average cosmic gravitational potential,* in accordance with

$$\bar{\phi} = -\tfrac{1}{3}c^2$$

Introduction

With this result, Mach furnishes a precise expression for the physical content of inertia-free gravitodynamics while at the same time making it possible to deduce, from gravitodynamics, the local relativistic physics (Treder, 1972).

It is clear then that Mach's principle occupies a key position in the discussion of relativistic theories of gravitation, inasmuch as it provides the bridge between their local and global aspects. As Einstein (1917) assumed at the time, it proves to constitute the third fundamental principle, next to the equivalence and general relativity principles, for the construction of general relativistic theories of gravitation. On the one hand, Mach's principle renders more concrete and less forbidding the radical content of inertia-free gravitodynamics in the sense of the local validity of the general theory of relativity. On the other hand, through the postulate of the unique determination of the world geometry by matter, it amplifies the assertions of the general theory of relativity, by imparting a more precise meaning to the effect due to cosmic masses upon space–time structure as understood from Einstein's equations.

Chapter 1

Local Principles and the Theory of Gravitation

1.1. Preliminary Ideas

Because of the weakness of gravitational forces, only a very limited number of experiments are suitable for the verification of any given theory of gravitation. Thus the problem of broadening the experimental foundation of the theory of gravitation is today at the center of gravitational research.

To make progress in this area of physics, it is not sufficient simply to catalog all the theories of gravitation that have been proposed to date and then to compare their experimental predictions with available experimental or observational data, such as those connected with celestial mechanical and optical effects. The main objections to a procedure of this kind are the following:

(1) It takes for granted that all conceivable theories of gravitation are already formulated explicitly or, if not, that they resemble known theories to the extent that they are ef-

fectively already contained in the aforementioned catalog of theories.

(2) Experiments and observations of sufficient accuracy that are feasible now or will be in the near future are hardly numerous enough to distinguish empirically between the theories in existence today.

(3) Moreover, as a matter of principle, any worthwhile theory should not confine itself to the data of experiment alone; it ought always to involve a large variety of assertions and predictions as well. For, if all that is desired were the mere explanation of the results of some actual experiments or observations, then that could in general already be accomplished simply by the inclusion of suitable *ad hoc* terms in the Lagrangian of the system under consideration.

An indifferent comparison of theory with experiment involves a simplification of the complex relation that exists between theory and experiment; in particular, such a procedure underestimates the possibilities for the establishment of new theories. But the catalog-and-compare method becomes a physically sensible methical tool if it catalogs theories primarily not in reference to any observable parameters, but with regard to some fundamental physical principles. Accordingly, in each instance, individual theories of gravitation become models of a class of theories which satisfy a certain combination of these principles.

It follows that experimentally verified predictions of any particular formulation of gravitational theory can be judged in the light of their significance for demonstrating the power of these fundamental principles. Such a procedure makes possible the discovery of basic experiments, whose results are capable of telling us if the answers that have been given in the form of physical principles to certain physical questions are right or wrong. These basic experiments constitute checks on predictions which Born (following Kant) called *synthetic predictions*. Only the assumption of principles or axioms (*synthetic axioms*) leading to such predictions enables us to formulate theories. Of course, the real meaning of these constructs will become clear only within the framework of the theories themselves. (Only the theories themselves can tell which entities are measurable, Einstein told Heisenberg.)

It is now necessary to distinguish essentially between two kinds of physical principles, which we shall call *correspondence principles* and *theoretical principles*.

By a correspondence principle we understand some postulate claiming that there exist special cases or limiting situations of the theory being formulated which are already contained within the old theory that is being replaced.

We designate as a theoretical principle any mathematically constructive statements or assumptions, expressing theoretical generalizations of certain physical experiences, which enable us to formulate new theories. Such principles can derive from experimentation and observation, past and present, in parts or all of physics or from theoretical approaches and theories that have withstood the test of time in physics. As for the nature of the theoretical generalizations, there exist different possibilities, because any theoretical extension (e.g., of the results of an experiment) already contains the beginnings of the speculation that is necessary for the formulation of a new theory. In regard to the formulation of the theory of gravitation, the most essential theoretical principles are the different formulations of the *principles of equivalence* and *relativity*.

Before discussing these principles and their meaning for the theory of gravitation, we have to choose between two different, but both conceivable, viewpoints: The theoretical principles can be formulated as applying either to infinitesimal space–time regions or to the entire universe. That the formulation for *finite* space–time regions is impossible, and thus not in need of consideration, follows from the universal character of gravitation: Gravitational fields, as is well known, cannot be screened; they can only be locally canceled. This problem was discussed in detail by Treder (1970) with reference to the relativity principle.

We start from the universal proportionality of Newton's *inertial* and (*passive*) *heavy masses*, m_I and m_P, respectively, which has empirically been verified to high (1 part in 10^{11} or better) accuracy for macroscopic bodies in homogeneous gravitational fields, and consider the consequences that follow from a choice between the above alternatives.

Einstein took the aforementioned proportionality—which, it must be remembered is confirmed only experimentally, and therefore approximately, although with extreme accuracy—and elevated it to a principle asserting the exact universal proportionality or equality of m_I and m_P for macroscopic objects in homogeneous gravitational fields. The *local equivalence between inertia and (passive) gravitation*, postulated in this way for macroscopic bodies, constitutes the *weakest formulation of the equivalence principle*. From it stems the idea for tackling the formulation of a new theory of gravitation by means of a unitary description of (passive) gravitation and inertia.

From our experience in the area of special relativity theory, we are familiar with a local description of inertia: Inertial forces can be described by a symmetric metric field g_{ik} with ten components. On the other hand, we know from the Newtonian theory of gravitation only the (instantaneous, action-at-a-distance) description of gravitation that defines a scalar potential in three-dimensional space. (This description, we may note, is a *global* one, in the sense that it assigns only secondary importance to local, field-theoretic, concepts, according to which the interaction between two bodies A and B is explained entirely in terms of the physical field values in the immediate vicinity of A and B, respectively.) Consequently, an attempt to allow for the connection between inertia and gravitation (as indicated by the equivalence between m_I and m_P), by a unitary local description of inertia and gravitation, brings about the reduction of gravitation to inertia. Conversely, a unitary global description of gravitation and inertia causes the reduction of inertia to gravitation (the Mach–Einstein doctrine; see Chapter 3).

These two procedures, the one *local* and the other *global*, for generalizing the equivalence principle for the inertial and heavy masses—which results, by virtue of the correspondence principle, in the reduction of gravitation to inertia, and *vice versa*—differ from one another in that they assign inertia and gravitation to different parts of the so-called *Poincaré sum*. Poincaré asserted that in physics neither the geometrical structure of space (G) nor any law of physical interaction (P) had any meaning when taken separately. Rather, the content of physics that is capable of being experienced is given by the Poincaré sum (Poincaré, 1912, 1914; Einstein, 1921)

$$G + P$$

From the standpoint of this sum, the Galilei–Newton inertia occupies an intermediate position [this was criticized already by Hertz (1894)], because the inertial forces predominate and belong to the part G of the Poincaré sum, while gravity in Newton's theory is a genuine force and accordingly belongs to part P. By virtue of the equivalence principle, however, inertial and gravitational forces must be lumped together. In accordance with our earlier remarks, the global approach classifies both inertial and gravitational phenomena under P in the sum $G + P$, while the local viewpoint assigns them both to the category G.

In the same way that our particular preference for either the local or global viewpoints determines the direction of generalization of the principle of equivalence of m_I and m_P, so this preference affects also the different generalizations of the other principles occurring in the formulation of gravitational theory. Here an intimate connection is found to exist between the different formulations of the equivalence principle and the statement of the relativity principle which is quite remarkable. This connection is shown on the one hand by the fact that, with respect to gravitation, only the validity of the equivalence principle allows the existence of a principle of relativity that is more general than that of Galilei or of Lorentz–Einstein. On the other hand, the connection becomes evident in the circumstance that different theories of gravitation, flowing from different formulations of the equivalence principle (in addition to some further principles), concord with the various formulations of the relativity principle.

[We add, parenthetically, that the first-mentioned connection is discussed for strictly local considerations in (Treder, 1970). It obtains also for the instantaneous action-at-a-distance description discussed in Chapter 3: Only the validity of the Mach–Einstein doctrine, which can be viewed as a global formulation of the equivalence principle, allows the generalization of the Galilei group.]

Turning now to the consideration of local relativity principles, we begin by discussing the *Lorentz–Poincaré group*. In special relativity theory, the Lorentz–Poincaré group (Latin indices, lower case and capitals, run from 0 to 3)

$$\bar{x}^i = \omega^i{}_l x^l + a^i \tag{1.1.1}$$

where

$$\partial_k \omega^i{}_l = 0$$
$$\partial_k a^i = 0 \tag{1.1.2}$$

and

$$\omega^i{}_l \omega_k{}^l = \delta^i{}_k \tag{1.1.3}$$

has (intentionally) three different interpretations:

(1) It represents the group of all orthogonal coordinate transformations in Minkowskian space–time.

(2) Alternatively, it signifies the isometry group of Minkowskian space–time.

(3) The homogeneous Lorentz group describes the transition from one inertial reference system Σ to another, Σ':

$$h'^A{}_i = \omega^A{}_B h^B{}_i \tag{1.1.4}$$

with

$$\omega^A{}_B \omega_C{}^B = \delta^1{}_C$$
$$h^A{}_i h_{Al} = \eta_{il} \tag{1.1.5}$$

Here

$$(\eta_{ik}) = \text{diag}(1, -1, -1, -1) \tag{1.1.6}$$

is the Minkowski metric tensor and the $h^A{}_i$ and $h'^A{}_i$ are *tetrads* defining the systems Σ and Σ', respectively; they can be constructed at each point of space-time with the help of three standard rods and one standard clock. (For further clarification of these and subsequent mathematical concepts, see Appendix B.)

Within the scope of its first meaning, the Lorentz–Poincaré group describes the passive transformation of the coordinates of a physical object. The second usage of the Lorentz–Poincaré group concerns the Helmholtz–Lie symmetries of the flat space–time manifold; thus it characterizes the possible active displacements of a body (translations and rotations). The equivalence between the two usages implies the maximal free mobility of objects in the nonstructured space–time of the Minkowski world.

Since the homogeneous Lorentz group also describes the transition from one inertial system to another, Lorentz covariance expresses Einstein's special relativity principle, which postulates the equivalence of all inertial systems for the description of physical phenomena. The (extensional) identity of Einstein's special relativity with the invariance under the group of the orthogonal centro-affine coordinate transformations in space–time allows us to identify the Cartesian space–time coordinates with the reference tetrads. Accordingly, the inertial reference systems take the form $h^A{}_i = \delta^A{}_i$, such that the transition from Σ to Σ' may be compensated by the corresponding contragredient coordinate transformation

$$\bar{h}'^B{}_i = \omega^B{}_A \omega^l{}_i \delta^A{}_l$$

$$= \delta^B{}_i \qquad (1.1.7)$$

In the local theory of gravitation and especially in the general theory of relativity, the Lorentz–Poincaré group is generalized in different ways, corresponding to its different interpretations. Here the Einstein group of general coordinate transformations

$$\bar{x}^i = \bar{x}^i(x^k)$$

$$d\bar{x}^k = \frac{\partial \bar{x}^k}{\partial x^i}\, dx^i$$

(1.1.8)

takes the place of the group of linear coordinate transformations.

The presence of a gravitational field imparts a structure to space–time. In general relativity theory, the metric becomes a Riemannian metric, as does the metric effectively acting on matter in the *bimetric* theories. Therefore, active displacements of bodies are generally no longer possible.

Einstein's general relativity extends special relativity by replacing the homogeneous Lorentz group, with constant coefficients $\omega^A{}_B$, by the corresponding group with coordinate-dependent coefficients $\omega^A{}_B(x^l)$. We then have

$$h'^A{}_i = \omega^A{}_B h^B{}_i$$

(1.1.9)

where

$$\omega^A{}_B = \omega^A{}_B(x^l)$$

$$\omega^A{}_B \omega_C{}^B = \delta^A{}_C$$

(1.1.10)

and

$$g_{ik} = h'^A{}_i h'_{Ak}$$

$$= h^A{}_i h_{Ak}$$

(1.1.11)

The foregoing defines a group of anholonomic transformations

29

that describes the local transition from one reference system to another.

Einstein's *covariance principle* requires the covariance of all equations of physics under the Einstein group (1.1.8), and his *principle of general relativity* their invariance under transformations of the local Lorentz group (1.1.9). In general relativity theory, both these principles are equivalent, as a consequence of Einstein's strong principle of equivalence (see below). [On account of this, the reference tetrads $h^A{}_i$ can be interpreted as *anholonomic space-time coordinates*. Weyl's lemma is valid; cf. Einstein and Mayer (1932).] In more general theories of gravitation, where such duality no longer exists, the principle of general relativity is violated. Under these circumstances, the nature of the gravitational field prescribes certain classes of reference systems. (Space–time is "teleparallelized," as proposed by Einstein; see below.)

According to Weyl (1923, 1924), the Einstein group $\bar{x}^i = \bar{x}^i(x^l)$ can be viewed as a generalization of the translational part a^i of the Lorentz–Poincaré group. It describes what happens when an observer replaces his reference system by another, i.e., when he changes his state of motion.

By contrast, the local Lorentz transformations with $w^A{}_B = \omega^A{}_B(x)$ are a generalization of the homogeneous part of the Lorentz–Poincaré group. They tell what the physical processes that are given in a global reference system look like in the proper rest frames of the single parts of the physical system under consideration which are undergoing arbitrary motions relative to one another; cf. Weyl (1929a, 1929b) and Treder (1971b).

Einstein's principle of relativity is a covariance principle that relates to strictly local operations. It is therefore a genuine field-theoretical principle: All physical objects are referred to the metric fields g_{ik} immediately surrounding them. On the other hand, since the time of Huygens, formulations of the relativity principle have existed that relate the dynamical properties of particles to the motion of distant masses. Such principles are important for global formulations of the theory of gravitation. They underlie Mach's principle of the relativity of dynamics, Poincaré's postulates of relativity, and the Mach–Einstein doctrine (cf. Chapter 3).

1.2. Linear Form of the Theory of Gravitation

If we start with a local consideration of gravitation, that is, with a study of gravitation in infinitesimal space–time regions, then the correspondence principle is seen to imply that any field theory of gravitation must be formulated so as to obey the following requirements:

(1) In spaces with vanishing gravitation, special relativity theory (especially its kinematics, electrodynamics, and optics) is strictly valid.

(2) In the limit of weak local gravitational fields and small relative velocities, Newton's theory of gravitation is approximately valid, that is, for

$$\frac{v^2}{c^2} \ll 1$$

$$\frac{fM}{rc^2} \ll 1$$

(1.2.1)

Newtonian gravitodynamics is the first-order approximation to the theory of gravitation, correct to terms of the order v^2/c^2. In equation (1.2.1), v denotes the relative speed, c light speed in vacuum, f

the Newtonian gravitational constant, M the mass of the gravitational field source, and r the distance between the field point and the center of the mass M. (In Chapters 1 and 2 we write the active heavy mass as M or m_A.)

In addition to the correspondence principle and the principle of the equivalence between inertial and passive heavy masses, the *principle of relativity* suggests itself as necessary for the formulation of a theory of gravitation. The concrete form of the relativity principle is, however, not determined *a priori*; it depends upon the structure of the laws of nature. As long as a gravitational law has not yet been formulated, the postulate of the validity of a concrete principle of relativity constitutes a restriction on the choice of the gravitational equations.

To begin with, we postulate the validity of the *special principle of relativity*, which, by the first correspondence principle requirement above, follows only in the case of vanishing gravitational fields.

The central physical idea for the formulation of the theory of gravitation derives from the theoretical generalizations of the principle of equivalence between the masses m_I and m_P. Thus, the first consequence of the correspondence principle gives rise to the following special-relativistic generalization, called *Einstein's weak principle of equivalence*:

In a laboratory that is freely falling in a gravitational field, special relativity theory is valid. Accordingly, inertial and gravitational forces cannot be distinguished by the interactions described in special relativity theory. The relative deviations from the predictions of special relativity theory are of the order

$$\frac{r^2}{R^2} \qquad (1.2.2)$$

Local Principles and the Theory of Gravitation

where r characterizes the linear dimensions of the laboratory and R denotes the corresponding local radius of curvature of the gravitational field. This result is a consequence of the so-called tidal deviation.

The foregoing Einsteinian formulation of the weak equivalence principle postulates the local validity of the special-relativistic equations in their canonical form, which contains first derivatives only of the physical variables.

From Einstein's weak equivalence principle and the special principle of relativity, it follows that the equations of motion for a structureless mass point are given by the familiar geodesic equation (generalized Galilean law of inertia)

$$m_I \frac{d^2 x^i}{d\tau^2} = -m_I \Gamma^i_{kl} \frac{dx^k}{d\tau} \frac{dx^l}{d\tau} \qquad (1.2.3)$$

which reduces for vanishing gravitation to the formula defining straight lines in Minkowski space. According to the equivalence principle, the right-hand term

$$-m_I \Gamma^i_{kl} \frac{dx^k}{d\tau} \frac{dx^l}{d\tau} = -m_P \Gamma^i_{kl} \frac{dx^k}{d\tau} \frac{dx^l}{d\tau} \qquad (1.2.4)$$

contains both the inertial and gravitational forces: $m = m_I = m_P$ represents both the inertial and the passive heavy masses of the particle.

The weak equivalence principle justifies the neglect of gravitation in elementary particle physics. All scalar conservation laws of the special-relativistic field theory are valid, too, if gravitational fields are present. Indeed, since the acceleration of a physical system (as a continuous procedure) cannot change its quantum numbers (particle numbers, charges, etc.) (Planck, 1907), neither therefore can a gravitational field. On the other hand, tidal influences cause the violation of the vectorial conservation laws for angular momentum and energy-momentum. These conservation laws reflect the Lie symmetries of the flat metric η_{ik}; the gravitational tensor generally does not possess these symmetries.

According to the weak equivalence principle and the special principle of relativity, the gravitational field is a space–time tensor G_{ik} (a tensor under Lorentz transformations). Theories of gravitation can differ from one another in that, either the G_{ik} are algebraically reducible to other more primary field functions (and then the field equations should be formulated for the primary entities), or the G_{ik} need to be supplemented by other primary entities describing the gravitational field (in which case, one must formulate field equations for the additional fields, too).

The possible algebraic constructions are the following:

(1)   A *scalar theory* results if one sets

$$G_{kl} = \psi \eta_{kl} \qquad (1.2.5)$$

with η_{kl} denoting the Minkowski tensor. The gravitational field is essentially the scalar field ψ.

(2)   A *tensor theory* can be based on the choice

$$G_{kl} = g_{kl} = \eta_{kl} + \gamma_{kl} \qquad (1.2.6)$$

The primary gravitational entity here is the arbitrary tensor γ_{kl} with ten components.

(3)   A *spin-vector theory* obtains if one postulates

$$G_{kl} = \eta_{AB} h^A{}_k h^B{}_l \qquad (1.2.7)$$

The tensor G_{kl} is generated by four reference tetrads $h^A{}_k$ (and spin vectors). The objects $h^A{}_k$ are viewed as primary gravitational potentials instead of the G_{kl}. The gravitational potential now possesses not ten but 16 components. The gravitational properties are always referred to the special-relativistic space–time. Concretely, they represent either necessary conditions for the functional dependence of the gravitational potential upon its sources (*potential representation*) or necessary conditions for the differential form of the gravitational field equations.

Accordingly, the consideration of Einstein's weak equivalence principle—deriving, firstly, by generalization from the equivalence between m_I and m_P and, secondly, from the first requirement of the correspondence principle—together with the special principle of relativity leads to assertions about the effect of gravitation on nongravitational matter.

In order to discover restrictions on the form of the equations determining the gravitational field, one has to take into account the second implication of the correspondence principle, requiring the approximate validity of Newtonian gravitodynamics for weak fields and low velocities. Because of Einstein's weak equivalence principle, and especially the associated geodesic law of motion for point masses, this postulate is guaranteed if the time–time component G_{00} of the first-order approximation is identified with the Newtonian gravitational potential $\phi = -fM/r$, in accordance with

$$G_{00} = 1 + \frac{2}{c^2}\,\phi \qquad (\phi < 0) \qquad (1.2.8)$$

From this equation, G_{00} is seen to satisfy the Poisson equation

$$\Delta G_{00} = \frac{4\pi}{c^2}\,f\varrho \qquad (1.2.9)$$

In the Newtonian limit, the mass density ϱ, which is special-relativistically the time–time component of the matter tensor, can be represented by

$$\varrho c^2 = \alpha T_0{}^0 + \beta T$$

$$\alpha + \beta = 1 \qquad (1.2.10)$$

(In the present section, $T_i{}^k$ will designate the special-relativistic energy-momentum tensor, and $T \equiv T_l{}^l = \eta^{lk}T_{lk}$.) The Poisson equation consequently becomes

$$\Delta G_{00} = \frac{8\pi f}{c^4}\,(\alpha T_0{}^0 + \beta T)$$

$$\alpha + \beta = 1 \qquad (1.2.11)$$

Starting from the consideration that both G_{kl} and T_{kl} are space–time tensors, one then sees that the special relativity principle requires the following special-relativistic formulation of equation (1.2.11) (Poincaré, 1906; Minkowski, 1908, 1909; Einstein, 1913; Einstein and Grossmann, 1913):

$$\Box \, G_{kl} = -\frac{8\pi f}{c^4}\,(\alpha T_{kl} + \beta \eta_{kl} T)$$

$$\alpha + \beta = 1 \tag{1.2.12}$$

From its derivation, this special-relativistic formulation is valid for arbitrary velocities, but only under the assumption that the gravitational potentials are small:

$$\frac{fM}{rc^2} \ll 1 \tag{1.2.13}$$

A generalization of the foregoing becomes possible if one introduces, over and above the tensor fields η_{kl}, G_{kl}, and T_{kl}, a scalar field, describing a source of gravitation which is unknown in special relativity theory. In the resulting scalar–tensor theory, a tensor function is added to the right-hand side of equation (1.2.12).

In equation (1.2.12), the constant $\beta = 1 - \alpha$ is still to be determined. Assuming that the gravitational potential is scalar (scalar theory), one obtains $\alpha = 0$ and $\beta = 1$. Then equation (1.2.12) with $G_{kl} = \psi \eta_{kl}$ describes the linear approximation of *Nordström's theory of gravitation* (Nordström, 1913; Einstein and Fokker, 1914), and Newton's action–reaction axiom is in general violated (von Laue, 1917). Abraham (cf. Abraham, 1920) was the first to prove the incompatibility of scalar theory (Nordström, 1913; Mie, 1923) with Newton's action–reaction axiom for the case of relativistic matter.

Consequently, if one demands the identity of m_A and m_P as an elementary form of the action–reaction principle, then a tensor theory has to be formulated, and the determination of the constant α follows from the principles formulated up to this point, without any additional assumptions:

Because of the validity of the axiom $m_A = m_P$, both in Newtonian gravitodynamics and (for completely static systems) in special relativity theory, it follows from the weak principle of equivalence that

$$m_I = m_P = m_A \qquad (1.2.14)$$

m_A being the source strength of the gravitational field. For isolated matter it furthermore ensues, from the weak principle of equivalence and the special relativity principle, that the inertial mass of a completely static system (relative to its rest system) is given in linear approximation by (Greek indices range from 1 to 3)

$$m_I = \frac{1}{c^2} \int (T_0{}^0 - T_\nu{}^\nu)\, dV \qquad (1.2.15)$$

[laws of von Laue (1917), Tolman (1934), and Einstein and Pauli (1943)]. In the Newtonian limit, this mass must be the active gravitational mass, that is, by equation (1.2.9),

$$m_A = \frac{c^2}{8\pi f} \int \Delta G_{00}\, dV \qquad (1.2.16)$$

Therefore, from equation (1.2.11),

$$\alpha = 2$$
$$\beta = -1 \qquad (1.2.17)$$

Local Principles and the Theory of Gravitation

Any choice for α other than the foregoing leads to the conclusion that the active heavy mass of a static system differs from the inertial and passive heavy masses by an amount of the order

$$(\alpha - 2)M \propto \frac{(\alpha - 2)}{2} \int T_0{}^0 \, dV \qquad (1.2.18)$$

In general, we obtain a deviation from Newton's action–reaction axiom $m_A = m_P$, and the conservation of the center-of-mass motion is no longer valid (see also below). The reason for this is that the active mass forms a static system only for $\alpha = 1 - \beta = 2$. And in the Newtonian limit, the condition $\alpha = 2$ is sufficient [cf. Treder (1973) and Jánossy and Treder (1972)].

The determination of the constant α can also be carried out as follows: The action–reaction axiom $m_A = m_P$ is an assertion that arises in the Newtonian limit from Einstein's *strong equivalence principle* (see Section 1.3). The requirement of the strict validity of this principle leads to Einstein's equations of general relativity theory, which, when combined with $G_{kl} = \eta_{kl} + \gamma_{kl}$, imply in the linear limit that

$$\frac{1}{2} (\Box \, \gamma_{kl} + \gamma_{,kl} - \gamma_k{}^{,r} - \gamma_{lr,k}{}^{r}) = - \frac{8\pi f}{c^4} \left(T_{kl} - \frac{1}{2} \eta_{kl} T \right)$$

$$(1.2.19)$$

where

$$\gamma = \eta^{ik} \gamma_{ik}$$

On imposition of *Hilbert's gauge condition* (Hilbert, 1915),

$$(\gamma_{kl} - \tfrac{1}{2}\eta_{kl}\gamma)_{,l} = 0 \qquad (1.2.20)$$

the left-hand term takes the form $\tfrac{1}{2} \Box \, \gamma_{kl}$. If the strong equivalence principle is to be satisfied in the linear approximation, the gravitational equations must become identical with the linearized Einstein equations (1.2.19) plus the Hilbert condition (1.2.20). (Obviously, it is sufficient to require that this be true for the Newtonian time component.)

Therefore, because of the special relativity principle, the weak equivalence principle, and the approximate validity of Newtonian mechanics (inclusive of the strictly satisfied action–reaction principle), the linear field equations have the form

$$\frac{1}{2} \, \Box \, \gamma_{kl} = - \frac{8\pi f}{c^4} \left(T_{kl} - \frac{1}{2} \, \eta_{kl} T \right) \qquad (1.2.21)$$

The aforementioned principles determine the linear form of the gravitational equations uniquely, as long as no sources unknown in special relativity theory are introduced.

The theory of gravitation with the field equations (1.2.21) that has been discussed up till now [Einstein, Abraham, Lorentz, Minkowski, Nordström, Planck, Mie, and especially Poincaré have investigated such a theory; cf. Abraham (1920) for the early terms of relativistic gravitational theories] is in many respects superior to the Newtonian theory, as we now indicate.

(1) For $v^2/c^2 \ll 1$, the special-relativistic theory of gravitation that is given by equation (1.2.21) contains Newtonian gravitodynamics (in the desired approximation), the Einsteinian clock dilatation, and, for $v \approx c$, Einstein's geometrical gravito-optics. The gravito-optics has no Newtonian limit. Both the Newtonian approximation and the gravito-optics for vanishing gravitational fields, i.e., for $G_{kl} \to \eta_{kl}$, tend to the special-relativistic results without gravitation.

The assertions of the relativistic gravito-optics qualitatively contradict the Newtonian laws when the high-speed limit $v \approx c$ is approached. The effect of the Newtonian gravitational potential ϕ on the motion of a particle coming from infinity is described in Newtonian dynamics by the deflection

$$\delta \chi = \frac{2 \mid \phi \mid}{\bar{c}^2} \tag{1.2.22}$$

and the speed change

$$v_{\text{light}} \approx \bar{c}\left(1 + \frac{\mid \phi \mid}{\bar{c}^2}\right) > \bar{c} \tag{1.2.23}$$

where $\bar{c}$ is the vacuum speed of light in the absence of gravitation $(\bar{c} = c)$.

If we restrict ourselves to the linear approximation, i.e., consider only terms of the order of the Newtonian gravitational potential in the functions G_{ik}, then we may write

$$ds^2 = \left(1 - \frac{2a}{r}\right) c^2 \, dt^2 - \left(1 + \frac{2b}{r}\right) dl^2 \tag{1.2.24}$$

where a and b are arbitrary constants and

$$dl^2 = (dx^1)^2 + (dx^2)^2 + (dx^3)^2 \tag{1.2.25}$$

In this approximation, we relativistically get

$$\delta \chi = \frac{4 \mid \phi \mid}{c^2}$$

$$= \frac{2}{r_0} (a + b) \tag{1.2.26}$$

$$a = b = fMc^{-2}$$

where r_0 is the periastron and

$$v_{\text{light}} = c\left(1 - \frac{2\,|\,\phi\,|}{c^2}\right)$$

$$= c\left(1 - \frac{a+b}{r}\right) < c \qquad (1.2.27)$$

In general, in the relativistic theory, for particles with speed $v_\infty = c$ at infinity, we have the speed law

$$v = \left(1 - \frac{a+b}{r}\right)v_\infty \qquad (1.2.28)$$

but in Newtonian theory we have instead the law

$$v = \left(1 + \frac{fm}{rv^2}\right)v_\infty \qquad (1.2.29)$$

Newtonian mechanics predicts that a particle is accelerated for all velocities of the particle, whereas relativistic gravitodynamics asserts that a particle with speed

$$v > \frac{c}{3^{1/2}} \qquad (1.2.30)$$

is slowed down by a gravitational field (Hilbert, 1916).

Using the metric (1.2.24) to calculate, via the geodesic equations, the motion of test particles with nonzero rest mass ("planets"), one gets the Kepler motion of planets as in Newton's theory; similarly, calculations for particles with vanishing rest mass (photons) lead to the gravito-optic effects discussed above. (See also Section 2.2.)

Local Principles and the Theory of Gravitation

Accordingly, Newton's theory of gravitation and (linear) gravito-optics hold to the same order in the gravitational constant f. Because of the relativity and equivalence principles, they form a mathematical unity. [Here a Newtonian particle can become a particle with a superrelativistic velocity, and conversely, by a proper change of the reference system (choice of Lorentz transformation).]

Together with the special theory of relativity, red shift and light deflection already determine the linearized gravitational equations.

(2) The (linear) special-relativistic theory of gravitation makes assertions about gravitational effects of higher order. These so-called *tidal effects* are determined by the curvature belonging to the gravitational potential, i.e., by the Riemannian tensor

$$R^i{}_{klm}[g_{st}] \tag{1.2.31}$$

They also give rise to *solenoidal effects* of the gravitational potentials: A directed entity Q changes its orientation, and therefore the magnitude Q^B, if it is transported n times along a closed circle of radius r, surrounding a gravity center of mass M at $r = 0$, in accordance with

$$\frac{\Delta Q^B}{Q^B} \approx n\,\frac{fM}{c^2 r}$$

$$\approx R^i{}_{klm}\,dF^{lm} \tag{1.2.32}$$

(de Sitter, 1916; Fokker, 1920, 1965; Weyl, 1923; Treder, 1971a). The solenoidal effect owes its occurrence to the form of the gravitational force

$$\sim \Gamma^i_{kl} \frac{dx^k}{d\tau} \frac{dx^l}{d\tau}$$

which is a direct consequence of the equivalence principle. Its existence is independent of the question whether the gravitational potential $G_{kl} = \eta_{kl} + \gamma_{kl}$ is interpreted, after Einstein, as a space–time metric tensor or, in the sense of a bimetric theory, as a universal tensor field. Furthermore, in the equations of motion for particles moving with sufficiently large velocities, the solenoidal effects are given by terms that represent Lorentz-force-like influences on the particle motion; these effects stem from the $G_{0\nu}$ component of the gravitational potential. Because of the linearized gravitational equations, such potentials are produced only by matter that does not form a completely static system and is described by a matter tensor with components $T_{0\nu} \neq 0$.

(3) According to the linearized gravitational theory, quasistatic matter produces a quasistatic gravitational potential (parentheses around repeated indices signify that the Einstein summation convention does not apply):

$$G_{00} - 1 = \gamma_{00} = -2 \int \frac{fT_{00}}{c^4 r} \, dV = -\frac{2fM}{c^2 r} = -\frac{2a}{r}$$

$$(1.2.33)$$

$$G_{(\nu\nu)} + 1 = \gamma_{(\nu\nu)} = -2 \int \frac{fT_{00}}{c^4 r} \, dV = -\frac{2a}{r}$$

This form of the gravitational potential is a consequence of the strong equivalence principle, in the sense of the Einstein–Pauli theorem and von Laue's condition. In the absence of these theorems, the quasistatic gravitational potential could have the general form

$$\gamma_{00} = -\frac{\varkappa}{4\pi} \int (\alpha + \beta) \frac{T_{00}}{r} \, dV = -\frac{2a}{r}$$

$$\gamma_{(\nu\nu)} = \frac{\varkappa}{4\pi} \int \beta \frac{T_{00}}{r} \, dV = -\frac{2b}{r} \tag{1.2.34}$$

$$\varkappa = \frac{8\pi f}{c^4}$$

In the *quasi-Newtonian* case, for which $v \ll c$, the equation of motion depends only on G_{00} and, therefore, the Einstein–Pauli–von Laue theorems become meaningless. But in gravito-optics, these theorems play an essential role.

The effects of gravito-optics (light deflection and deceleration of photons in the gravitational field of the sun) were measured and found to be in agreement with the predictions of the Einstein–Pauli–von Laue theorems. This finding rules out scalar theories, which predict no influence of gravitation on the propagation of light; then $a = -b$. Neither Lorentz-force-like nor solenoidal effects could previously be measured; but today the sufficiently stabilized motion of satellites provides a method for measuring such effects with good accuracies.

We should add the remark that, in the case of sufficiently general matter, the gravitational potential contains a general three-dimensional tensor $-\delta_{\mu\nu} + \gamma_{\mu\nu}$. The tensor $\gamma_{\mu\nu}$ causes an isotropy of inertia, which, although arising from the linear approximation of the gravitational potential, in the equations of motion represents an effect of higher order.

(4) The validity of the special relativity principle requires the replacement of the Laplace operator Δ by the d'Alembert operator

$$\Box = \frac{1}{c^2} \frac{\partial^2}{\partial t^2} - \Delta$$

and, in this way, the transition from the potential equation to the wave equation (Poincaré). With the wave equation is connected a retardation of the gravitational actions, with a finite propagation speed c, and the possible existence of gravitational waves, especially of free gravitational waves, satisfying the equation (Einstein, 1918)

$$\Box G_{kl} = \Box \gamma_{kl} = 0 \qquad (1.2.35)$$

Emission and absorption of gravitational waves and, therefore, their experimental evidence are restricted by the equivalence principle. This principle, firstly, rules out the existence of negative heavy masses and, therefore, the existence of gravitational dipoles. Secondly, a mass that is freely moving in a gravitational field should not emit gravitational radiation, because free motion in a gravitational field is "force-free" according to the equivalence principle (Mie, 1923). Gravitational radiation is expected to be quadrupole in nature. Consequently, the simplest antenna having any use must be a quadrupole antenna. With such an antenna, the deviation, i.e., a tidelike effect, can be measured: The spatial variation of the gravitational field strength and, therefore, dynamically, the accelerations, are proportional to the Riemannian tensor $R^i{}_{klm}$. This is a consequence of the equivalence principle, according to which not the accelerations but the relative accelerations produce a physical effect: Since the motions in a gravitational field, on account of Einstein's weak equivalence principle, are "force-free," the motion of a charged particle does not generate any "acceleration waves."

In view of the validity domain of the equivalence principle, this conclusion is strictly valid only in homogeneous regions of the gravitational field. This means that, for free motions in a gravitational field, radiation corrections are proportional to the tensor R^i_{klm}, that is, proportional to the deviation, not the acceleration.

Despite the advantages of the special-relativistic theory of gravity over the Newtonian theory, there are compelling reasons for proceeding to a more complex (and, in general, nonlinear) theory of gravitation, relative to which the theory based on equation (1.2.21) can be interpreted to serve only as linear approximation. The reasons are the following:

(1) Newtonian celestial mechanics is empirically sound only up to errors of the order

$$\left(\frac{v^2}{c^2}\right)^2 \approx \left(\frac{fM}{rc^2}\right)^2$$

Accordingly, the undisturbed planetary orbits are not closed Kepler ellipses, as predicted by Newtonian mechanics; instead they are rosettes with a small perihelion rotation $\delta\psi$, which is furnished to a high degree of accuracy by the Einstein formula (Einstein, 1916b)

$$\delta\psi = 6\pi \frac{fM_\odot}{c^2} \frac{1}{r_0(1 - e^2)} \equiv 3\Omega \qquad (1.2.36)$$

wherein M_0 is the solar mass and $2r_0$ and e denote the major axis and eccentricity, respectively, of the ellipse under consideration.

The linear theory of gravitation, by contrast, predicts

$$\delta\psi = 4\Omega \qquad (1.2.37)$$

that is, a rotation in excess of the Einsteinian value. Of course, it should be mentioned that the perihelion motion also depends on the internal structure of the sun. But then this is already the case in the Newtonian limit if the mass distribution is not spherically symmetric but quadrupolelike. Depending on the nature of this quadrupole, the additional perihelion motion can have a positive or a negative sign. The quadrupole effects are given by

$$\frac{6\pi}{5} \frac{\Delta R}{R} \left(\frac{R}{r} \right)^2 \propto \frac{R\,\Delta R}{r^2} \qquad (1.2.38)$$

R being the equatorial radius of the sun and ΔR its excess over the polar radius. These effects can best be observed in the motions of Mercury and the asteroid Icarus.

On the other hand, in a general theory of gravitation, the structure of a spherically symmetric field depends upon the equation of state of the matter constituting the field source. [Because of Birkhoff's theorem (Birkhoff, 1923), this is not the case in Einstein's general relativity theory.] In general theories of gravitation, one has

$$a \propto \int (T_0{}^0 - T_\nu{}^\nu)\, dV \approx \int (c^2\varrho + 3p)\, dV \qquad (1.2.39)$$

and

$$b \propto \int \left(T_0{}^0 + \frac{1}{3} T_\nu{}^\nu \right) dV \approx \int (c^2\varrho - p)\, dV \qquad (1.2.40)$$

In the linear theory, the perihelion is found to advance through the angle

$$\delta\psi = \frac{2\pi}{r_0(1 - e^2)}(a + 2b) \qquad (1.2.41)$$

From $b < a$, a significant effect is to be expected only if an essential part of the solar mass is superrelativistically degenerated. But astrophysically this would be difficult to establish. With $a \neq b$ in celestial mechanics (and, generally, for multiple stars, too) one has to distinguish between the active heavy masses a and the inertial masses $b = fmc^{-2}$; therefore, the action–reaction axiom is violated. But methods of celestial mechanics determine the active heavy masses only, and here, for purely gravitational interactions, the mass-center motion of the active heavy masses M_k is conserved:

$$\sum_k M_k \ddot{x}_k^\nu = \sum_k M_k \ddot{x}^\nu = 0$$

$$\qquad (1.2.42)$$

$$x^\nu = \sum_k M_k x_k^\nu \left(\sum_k M_k\right)^{-1}$$

Consequently, $M \neq m_P$ is compatible with the Newtonian limit of celestial mechanics, as long as only gravitational forces act among the bodies. By contrast, $M \neq m_P$ fails to conserve the center-of-mass motion of inertial masses (Treder, 1974), i.e.,

$$\sum_k m_k \ddot{x}_k^\nu \neq 0 \qquad (1.2.43)$$

It should also be mentioned that, according to Anding (1905), von Seeliger (1906), and Aoiki (1967), it is uncertain whether the reference system that has been employed in astronomy until now is in fact a nonrotating inertial system.

(2) The linear theory of gravity leads to obvious nonsense if it is extended to extremely large masses M and mass densities ϱ. In the linear theory, the effects of masses add up linearly. Thus, for very large active masses of finite density the gravitational field strengths increase without limit:

$$M = m = c^{-2} \int (T_0{}^0 - T_\nu{}^\nu)\, dV$$

$$= \int \left(\varrho + 3\, \frac{p}{c^2} \right) dV \qquad (1.2.44)$$

In order to prevent the collapse of such mass distributions, the gravitational force $\sim Mr^{-2}$ must be compensated by a centrifugal force $\sim v^2 r^{-1}$. The virial law leads to the estimate

$$\frac{fa}{r} = \frac{fM}{rc^2} = \frac{v^2}{c^2} \approx \frac{f\varrho r^3}{c^2 r} \qquad (1.2.45)$$

from which we can draw the conclusion that fMc^{-2} is larger than r only for $v > c$. But in a special-relativistic theory, c is the maximum speed. Therefore, mass systems with $fMc^{-2} > r$, r being the radius of the system, must collapse indefinitely, "leaving space" when r becomes zero. Such a collapse is the final stage in the evolution of stars with mass not significantly larger than that of the sun; above all, such a stage occurs in the centers of galaxies.

From the foregoing, it follows that the linear theory of gravity, though originating in special relativity theory, has consequences that contradict the special theory. The linearized theory of gravitation, therefore, cannot be the correct theory of gravitation. (Unfortunately, similar difficulties appear in Einstein's general relativity theory; see below.)

Local Principles and the Theory of Gravitation

(3) The special-relativistic theory of gravitation of necessity uses two metrics (two "ethers"), viz., a Minkowski metric η_{ik}, with regard to which the gravitational equations are formulated, and a gravitational metric $G_{kl} = \eta_{kl} + \gamma_{kl}$, which is determined by solutions of the gravitational equations plus boundary conditions, relative to which, according to Einstein's weak equivalence principle, nongravitational physics is formulated. In principle, this implies a distinction between passive heavy mass and active mass. The strong principle of equivalence now requires that the gravitational equations also be formulated with reference to the metric G_{ik}, i.e., that only one "ether for gravitation and electrodynamics exists" (Hertz, Einstein, Weyl). This strong equivalence principle necessarily leads to a nonlinear theory of gravitation, that is to say, to the general theory of relativity.

(4) Gravitation is an extremely weak long-range interaction whose quanta propagate at light speed c, i.e., the quanta do not possess any rest mass. In elementary particle physics it is known that such interactions are connected with a breakdown of elementary physical symmetries. The only symmetry that is possible in the gravitational theory is the symmetry of Minkowski space, which is described by the Lorentz–Poincaré group of the special theory of relativity. This Lorentz covariance is valid in the special-relativistic tensor theory of gravitation, too. Hence there is no breakdown here that is compensated by gravitational fields.

(5) If the tensor theory of gravitation based on equation (1.2.21) is viewed as strictly valid, it leads to an anisotropy of the inertial masses that is given by the three-dimensional inertia tensor (Fokker, 1965)

$$m_{\mu\nu} = m\left[\delta_{\mu\nu}\left(1 + 3\frac{|\phi|}{c^2}\right) - \gamma_{\mu\nu}\right] \qquad (1.2.46)$$

The equivalence principle reduces possible anisotropy effects to those that can conceivably be measured in laboratories that are not freely falling. For a body resting on the Earth, the tensor gravitational potential of the Earth gives rise to a mass anisotropy corresponding to

$$\gamma_{\mu\nu} = -\frac{4}{3}\frac{f}{c^2}\frac{M}{r}\frac{V_\mu V_\nu}{c^2} \approx 2 \times 10^{-21} \qquad (1.2.47)$$

V_μ being the equatorial rotational speed of the Earth and M its mass. This kind of mass anisotropy, which should be measurable by the techniques of nuclear physics, does not seem to exist. (Today the possibility of a mass anisotropy in the general theory of relativity, too, is being raised; see Treder, 1973.)

(6) The linear theory implies a dependence of inertia upon the gravitational potential in accordance with

$$m_I = m_0\left(1 + \frac{3a}{r}\right) \qquad (1.2.48)$$

Accordingly, mass inertia, $m_I = m_0$, exists also in gravitation-free $(a = 0)$ spaces. The dynamical effects, as determined by the space structure that is independent of the masses (i.e., determined by the Minkowskian background η_{kl}), generally exceed the gravitational effects by many orders of magnitude. This is a direct consequence of the correspondence principle. [The special-relativistic theory of gravitation teaches the "predominance of ether over matter" (Weyl, 1924).]

(7) The vacuum field equations of the linear theory of gravitation can be derived from the Hamiltonian principle with the

Local Principles and the Theory of Gravitation

Lagrangian

$$\mathscr{L}_G = -\tfrac{1}{4}\eta^{ik}\eta^{lm}\eta^{uv}(g_{il,u}g_{km,v} - \tfrac{1}{2}g_{ik,u}g_{lm,v}) \qquad (1.2.49)$$

In special-relativistic field theories, therefore, the canonical *energy-momentum tensor*

$$t_i^{\ k} = \frac{\partial \mathscr{L}_G}{\partial g_{lm,i}}\, g_{lm,k} - \delta_i^{\ k}\mathscr{L}_G \qquad (1.2.50)$$

belongs to the gravitational field. If one neglects the backward action of the gravitational field on matter, that is, considers a purely special-relativistic matter tensor T_{ik}, then it is seen that the field equations in the presence of matter are derivable from the Lagrangian

$$\mathscr{L} = \mathscr{L}_G + \varkappa\mathscr{L}_{\text{mat}} \ (= \mathscr{L}_G + Kg^{ik}T_{ik}) \qquad (1.2.51)$$

But if one also takes into account the reaction of the gravitational field on matter and, accordingly, intends to derive the entire gravitodynamics from a closed Hamiltonian (where Einstein's weak equivalence principle must be strictly obeyed), then the special-relativistic field identities

$$\partial_i[\varkappa(-g)^{1/2}T_k^{\ i} + t_k^{\ i}] = 0 \qquad (1.2.52)$$

which follow from the Lorentz invariance of the Hamiltonian integral, must be identical with the dynamical equation

$$\varkappa(-g)^{1/2}T_i^{\ k}{}_{;k} = 0 \qquad (1.2.53)$$

which stems from Einstein's weak equivalence principle.

This important point was first established by Einstein (1913, 1914). His demonstration requires the transition to a nonlinear theory of gravitation, where the canonical energy-momentum tensor ℓ_i^k of the gravitational field itself acts as a source of gravitation. The genuinely special-relativistic gravitational equations thus become nonlinear and exhibit an Einstein (or Einstein–Rosen) form. The first formulation of such an equation was furnished by Einstein and Grossmann (1913) only in a Lorentz-covariant form, i.e., as a genuinely special-relativistic field equation. In keeping with the arguments about the linear approximation presented above, this equation must have the form

$$\square\,[(-g)^{1/2}g^{ik}] = -2[\varkappa T^{ik}(-g)^{1/2} + \ell^{ik}] \qquad (1.2.54)$$

(A discussion of this result will be given later.) Here, $(-g)^{1/2}g^{ik}$, by virtue of the field identities, obeys the *de Donder condition* (cf. Fock, 1960),

$$\partial_i[(-g)^{1/2}g^{ik}] = 0 \qquad (1.2.55)$$

which is a nonlinear generalization of the *Hilbert condition*. The Lagrangian

$$\mathscr{L} = \varkappa(-g)^{1/2}L_{\mathrm{mat}} + \mathscr{L}_G$$
$$= \varkappa(-g)^{1/2}g^{ik}T_{ik} + \mathscr{L}_G \qquad (1.2.56)$$

belonging to equation (1.2.54) then is *Einstein's Lagrangian of general relativity theory* or one of *Kohler's Lagrangians* characterizing a *bimetric* theory (see later).

1.3. Nonlinear Form of the Theory of Gravitation: Statement of the Problem and First Steps

The physical difficulties that were shown to appear in connection with the use of the linear gravitational equations make it imperative for us to search for generalizations of these equations or even pursue a completely new approach to the description of gravitation. Since the present chapter deals with the local aspects of gravitation, we will investigate here only the question of the generalization of the special-relativistic linear gravitational equations (1.2.21). An entirely different possibility for describing gravitation will be discussed in Chapter 3.

There is an abundance of conceivable (linear and nonlinear) generalizations of equations (1.2.21). If one adheres to the above special-relativistic formulations of the theoretical principles (relativity and equivalence principles), then it is not possible to go beyond the *linear* theory of gravitation, that is, to construct nonlinear generalizations of the gravitational theory founded on equations (1.2.21). On the other hand, these formulations of the theoretical principles do not exclude modifications of equations (1.2.21) that stay within the framework of the linear special-relativistic theory.

To rule out such modifications of the gravitational equations, we have to consider two further postulates. Unlike the large principles, these postulates do not result from direct generalization of certain physical experiences; rather, they are primarily *pragmatic* postulates. These new principles are as follows:

1. In addition to Newton's gravitational constant, no new constant having the dimension of a reciprocal area is introduced.

2.   The field equations of gravitation are differential equations of the second order.

Actually, in gravitational equations of higher order, one generally needs besides f and c further dimensional fundamental constants to ensure the dimensional equality of both sides of the inhomogeneous field equations.

If one should, contrary to the requirement (1), introduce new constants of the dimension cm^{-2}, then it must be assumed that the additional terms appearing in the linear field equations as a result of this procedure do not significantly modify celestial mechanics; they should be relevant only for cosmological solutions of the equations modified in this way (see below). There are two terms of this sort in the following field equations:

$$\Box\, G_{kl} - 2\lambda\eta_{kl} + k^2(G_{kl} - \eta_{kl}) = -2\varkappa T_{kl}^* \qquad (1.3.1)$$

where λ corresponds to Einstein's *cosmological constant* and k to *Seeliger's constant of absorption*. Under inclusion of these terms, the generalized Poisson equations read

$$-\varDelta\gamma_{00} - 2\lambda + k^2\gamma_{00} = -\varkappa\varrho c^2 \qquad (1.3.2)$$

Equation (1.3.2) has, for $k = 0$, the elementary solution of the Weyl–Trefftz type (Weyl, 1923):

$$\gamma_{00} = -\frac{2fM}{c^2 r} - \frac{\lambda}{3}r^2$$

$$(1.3.3)$$

$$M = \text{const}$$

and, for $\lambda = 0$, the elementary solution of the Seeliger–Neumann type:

$$\gamma_{00} = - \frac{2fM}{c^2 r} e^{-kr}$$

$$(1.3.4)$$

$$M = \text{const}$$

here we have chosen $k > 0$.

The introduction of the constants k and λ is empirically justifiable if either

$$\lambda \approx R^{-2} \qquad (1.3.5)$$

where R is the average curvature radius of the cosmic space, or

$$k \approx R^{-1} \qquad (1.3.6)$$

where R is an effective average radius of curvature. According to Yukawa and de Broglie, k (>0) can be interpreted as the rest mass μ of gravitons,

$$k = \frac{\mu c}{h} \qquad (1.3.7)$$

We note that a replacement of the linear second-order differential equation

$$\Delta G_{00} = \frac{4\pi f}{c^2} \varrho \qquad (1.3.8)$$

for instance, by the differential equation of the fourth order

$$\Delta^2 G_{00} = \frac{4\pi f}{c^2} \, \Delta \varrho \tag{1.3.9}$$

would generally represent a significant modification also in the case of celestial mechanics. Most importantly, it would lead to a new definition of the active heavy mass.

However, in view of the clear advantages of the linear theory formulated in Section 1.2, only a few arguments have been advanced so far for such modifications, especially because none of the new proposals are able to remove most of the shortcomings mentioned above. At the same time, one should not forget that there is a defect of the linear theory that can be remedied by modifications of this sort.

As we pointed out earlier, the linear gravitational equations (1.2.21) are not able to sensibly describe the behavior of very large masses or mass densities. According to this theory, spherical mass systems with radius $R > fMc^{-2}$ must collapse indefinitely until the total mass is concentrated in a single point. This singular final stage in the evolution of dense objects is in fact given mathematically by the singularity at $r = 0$ of the spherical static solution of equations (1.2.21).

Such singularities do not, however, occur in gravitational theories, where the spherical static solution

$$G_{00} = 1 - \frac{2fM}{c^2 r}$$

$$G_{0\nu} = 0 \tag{1.3.10}$$

$$G_{\mu\nu} = -\delta_{\mu\nu}\left(1 + \frac{2fM}{c^2 r}\right)$$

of equations (1.2.21), is replaced by a solution of the type

$$G_{00} = 1 - \frac{2fM}{rc^2} \left(1 - e^{-Kr} \right)$$

$$G_{0\nu} = 0 \qquad\qquad (1.3.11)$$

$$G_{\mu\nu} = -\delta_{\mu\nu}\left[1 + \frac{2fM}{c^2 r} \left(1 - e^{-Kr} \right) \right]$$

One now does obtain spherical static potentials of the latter form as solutions of the field equations that can be constructed in analogy with the fourth-order electrodynamics proposed by Bopp (1940) and Podolsky (1941); cf. Treder (1975). These gravitational field equations, which are partial differential equations of the fourth order, describe the unification of long-range gravitation [defined in the static spherical symmetric case by equation (1.3.10)] with the likewise universal, but short-range, interaction that corresponds to the Yukawa-like potential contribution $r^{-1}e^{-Kr}$ occurring in (1.3.11). In the gravitational theory constructed in this way, a fundamental constant appears which has the dimension of a reciprocal length and can be interpreted as the reciprocal Compton wavelength $K = mc/h$ of a "heavy graviton" of mass m. Accordingly, in this theory, two types of field quanta are assigned to the gravitational field, namely, the usual massless gravitons and the heavy gravitons (mesons of spin 2) belonging to the short-range field contribution.

It should, however, be stressed that this kind of gravitational theory, devised in analogy to the electrodynamic theory of Bopp and Podolsky, violates both of the two pragmatic principles given above: It involves differential equations of the fourth order and it contains a new fundamental constant K^2 having the dimension of a reciprocal area.

Apart from the possibility that it might solve the collapse and singularity problem, there are no significant arguments for the generalization of the field equations (1.2.21) that is here contemplated. We shall therefore, in what follows, give no further consideration to such modifications of the theory, and instead discuss nonlinear modifications of the linear theory (1.2.21), in agreement with both of the aforementioned pragmatic principles.

The first hint of the need for a nonlinear (with respect to the G_{ik}) modification of the linear equations (1.2.21) comes from the so-called *post-Newtonian approximation* of celestial mechanics. The use of this approximation means that, in the equations of motion, one considers, for bodies with relative speeds $v \ll c$, only terms to the order

$$\left(\frac{v^2}{c^2} \right)^2 \approx \left(\frac{fM}{rc^2} \right)^2 \tag{1.3.12}$$

Now, the generalization of the linear equations of gravitation can be carried out by means of the argument that the quantities γ_{kl}, appearing in $G_{kl} = \eta_{kl} + \gamma_{kl}$, are small in celestial mechanics,

$$|\gamma_{kl}| \approx \frac{fM}{c^2 r} \ll 1 \tag{1.3.13}$$

since they contain the Newtonian potential $\sim Mr^{-1}$ multiplied by the small parameter fc^{-2}. The linear equations (1.2.21) are then viewed as the first-order equations resulting from the expansion of the generalized equations in this small parameter (or the small parameter $v^2 c^{-2}$).

Local Principles and the Theory of Gravitation

For the problem discussed above, then, it is sufficient to know the first-order components $\gamma_{\mu\nu}$ and $\gamma_{\nu 0}$ of the potential (i.e., for these components, the consideration of the linearized equations is sufficient). But for the time-time component γ_{00}, a nonlinear generalization of the Poisson equation is permitted. Considering the linear approximation, the general form of the "nonlinear Poisson equation of second order" can then be determined, if one requires that it contains at most bilinear terms and that no new physical constants are introduced. The *generalized Poisson equation* then has the form

$$\Delta\gamma_{00} + \mu\eta^{kl}\gamma_{00,k}\gamma_{00,l} = \frac{8\pi f}{c^4}(1 + \nu\gamma_{00})T_{00}^* \qquad (1.3.14)$$

where

$$T_{ik}^* \equiv T_{ik} - \tfrac{1}{2}\eta_{ik}T \qquad (1.3.15)$$

and μ and ν are two numerical constants. Assuming an appropriate coefficient μ of the nonlinear term on the left-hand side of (1.3.14), one can give the perihelion motion any desired value. From the form

$$\frac{8\pi f}{c^4}(1 + \nu\gamma_{00})T_{00}^* = \frac{8\pi \tilde{f}}{c^4}T_{00}^* \qquad (1.3.16)$$

of the right-hand side, $\nu \neq 0$ is seen to imply that the effective gravitational constant $\tilde{f}$ depends on the Newtonian gravitational potential $\phi = \tfrac{1}{2}\gamma_{00} - 1$ itself. In order to discover the exact form of the nonlinear generalization of the special-relativistic equations of gravitation, one has to look for a Lorentz-covariant generalization of equation (1.3.14). Since there are several such generalizations, we restrict the possibilities by requiring: (1) The field equations should be derivable from a closed Lorentz-covariant Hamiltonian

61

principle (Einstein 1915, 1916a; Hilbert 1915) and (2) the Lagrangian must be bilinear with respect to $G_{ik,l}$.

These requirements lead to a variety of gravitational equations, which were considered by Kohler (1952, 1953) and von Laue (1956). According to Kohler, our choices are further limited by the postulate that (in analogy to general relativity theory and after Einstein and Hilbert) the normal-coordinate Lagrangian can be written in the form of a functional of the g_{ik} themselves. We demand finally that the gravitational field has the field-dynamical properties of a special-relativistic field, i.e., that the Lagrangian is a homogeneous bilinear function in the first derivatives of the g_{ik} and defines a symmetric canonical energy-momentum complex. One then gets

$$\mathscr{L} = \lambda_\mathrm{I} \mathscr{L}_\mathrm{I} + \lambda_\mathrm{II} \mathscr{L}_\mathrm{II} + \varkappa_0 \mathscr{L}_\mathrm{mat} \tag{1.3.17}$$

where

$$\mathscr{L}_\mathrm{I} = \frac{1}{(-g)^{1/2}} g_{mn} g_{rs} g^{lt} \mathscr{g}^{mr}{}_{,l} \mathscr{g}^{ns}{}_{,t} \tag{1.3.18}$$

and

$$\mathscr{L}_\mathrm{II} = \frac{1}{(-g)^{1/2}} g_{mn} g_{rs} g^{lt} \mathscr{g}^{mn}{}_{,l} \mathscr{g}^{rs}{}_{,t} \tag{1.3.19}$$

$$\mathscr{g}^{ik} \equiv g^{ik}(-g)^{1/2}$$

The connection of the foregoing with the linear theory can be established if one assumes that

$$\varkappa_0 = \lambda_\mathrm{I} \varkappa \tag{1.3.20}$$

and

$$\lambda_\mathrm{II} = -\tfrac{1}{2}\lambda_\mathrm{I} \tag{1.3.21}$$

In this manner the gravitational and numerical constants are ascertained.

With the choice of constants determined by the last two equations, Kohler's theory becomes identical with Rosen's interpretation of Einstein's general relativity theory (Rosen, 1940). This means that Kohler's equations can be written in normal coordinates as Einstein's equations

$$R_i{}^k = -(\varkappa T_i{}^k - \tfrac{1}{2}\delta_i{}^k T_{mn} g^{mn}) \qquad (1.3.22)$$

which, in harmonic coordinates, assume the form

$$\mathfrak{g}^{kl}{}_{,l} = 0 \qquad (1.3.23)$$

Cf. von Laue (1956).

The nonlinear special-relativistic generalizations that were considered by Kohler provide a whole class of field equations.

The principles and postulates discussed until now are incapable of uniquely determining the equations of gravitation. A stronger restriction and, moreover, a unique determination of the equations of gravitation result if we assume the validity of the large theoretical principles.

1.4. Theoretical Principles and Nonlinear Theory of Gravitation

According to the remarks of the previous section, the weak (special-relativistic) equivalence principle and the special principle of relativity are comparatively weak restrictions. However, for both these principles there exist more restrictive formulations. Recalling our introductory remarks (Section 1.1), two possibilities are available to us for generalizing the principle of relativity. They correspond to two possible choices of the functions describing the gravitational fields.

The formulation of a general principle of relativity, arising from the inhomogeneous part of the transformation (1.1.1) and requiring the covariance of physical equations with respect to arbitrary coordinate transformations, is primarily connected with purely space–time quantities. However, to describe the complex "gravity + inertia," one can, in addition to η_{kl} and G_{ik}, introduce further space–time functions, e.g., a scalar field. Accordingly, there are many ways in which to formulate a theory of gravitation that satisfies the covariance principle.

If one takes into account that, in addition to G_{ik} (and possibly further gravitational functions), there exists moreover a Minkowski background η_{kl}, then it becomes clear that the covariance of physical equations with respect to arbitrary coordinate transformations does not at all imply the validity of a general principle of relativity: Because of the presence of the background space, the Lorentz–Poincaré group remains the genuine symmetry group; thus, the special theory of relativity remains valid.

The unique determination of a system of gravitational equations is based on the requirement of the validity of Einstein's *strong* equivalence principle. According to this principle, gravitation is completely determined entirely by the functions G_{ik}; both the background metric η_{kl} and additional (e.g., scalar) fields are denied any necessary role. Consequently, the gravitational equations must be formulated with respect to the space–time metric $G_{ik} = g_{ik}$, i.e., relative to a geometry defined by the gravitational potential itself. In this manner, the coordinate covariance of the equations of gravitation becomes identical with Einstein's general relativity, for, the group of general coordinate transformations becomes identical with the Einstein group of gauge transformations of the field equations: One has ten field equations with four differential identities, and therefore the field equations must be derived from a closed Hamiltonian principle. Here the Lagrangian of the gravitational field is a functional of the g_{ik} alone. If one requires the Lagrangian to be bilinear in $g_{kl,m}$, then Einstein's Lagrangian is well determined and leads in a unique fashion to Einstein's equations

$$R_i{}^k = -\varkappa(T_i{}^k - \tfrac{1}{2}\delta_i{}^k T)$$
$$T = g^{mn}T_{mn}$$

(1.4.1)

(The cosmological form of the Einstein equations is also compatible with the general principle of relativity and the strong equivalence principle. Here the Weyl–Trefftz metric replaces the Schwarzschild metric.) According to general relativity, the relation

$$(R_i{}^k - \tfrac{1}{2}\delta_i{}^k R)_{;k} = -\varkappa T_i{}^k{}_{;k} = 0 \tag{1.4.2}$$

which is an expression of the gauge invariance of the field equations (1.4.1), is identically fulfilled. A result of this general relativity is that four of the ten g_{ik} are not determined by (1.4.1), which indetermination is an expression of the coordinate covariance; establishment of the four functions implies a choice of a coordinate system. If this coordinate system is fixed by the harmonic coordinate condition

$$[(-g)^{1/2}g^{ik}]_{,k} = 0 \tag{1.4.3}$$

then Einstein's general theory of relativity is extensionally identical with the *Rosen–Kohler* theory (Rosen, 1940; Kohler, 1952, 1953; von Laue, 1956); see above.

Einstein's general relativity theory furnishes the so-called *Einstein value* of the perihelion motion and, of course, the entire linear theory. Einstein's theory is distinguished by the fact that the exterior gravitational field belonging to a spherically symmetric body depends solely on one constant m; this constant characterizes all types of mass of the body—the inertial, active heavy, and passive heavy masses. In Einstein's general theory of relativity—and only in this theory and that of Rosen and Kohler—all three masses are identical and, in the case of a static gravitational field, given by the *Tolman–Einstein–Pauli value*,

$$m_I = m_A = m_P = \frac{2}{c^2} \int (-g)^{1/2}(T_0{}^0 - T_\nu{}^\nu)\, d^3x \qquad (1.4.4)$$

(This identity is just the dynamic expression of the strong equivalence principle, which in general relativity theory manifests itself in the Birkhoff theorem.)

Without going here into the possibilities for an experimental investigation of these masses, we can state that the empirical determination of whether these masses are equal or not constitutes the crucial test for Einstein's general relativity theory. If these masses should turn out to be identical, then general relativity theory (or the Rosen–Kohler theory) would be the only possible local theory of gravitation.

The formula (1.4.4) asserts that the active gravitational charge m_A of a body increases linearly with increasing inertial mass (i.e., with an increasing quantity of matter or particle number, respectively). Therefore, in the general theory of relativity (as in the linear theory) one has to expect a "gravitational collapse" of massive celestial objects. The exterior gravitational potential increases linearly with the increase of particle number.

In general relativity theory, the exterior spherically symmetric gravitational field is unambiguously given by the Schwarzschild metric, which in spherical coordinates reads

$$ds^2 = \left(1 - \frac{2fm}{rc^2}\right)c^2\,dt^2 - \frac{1}{1 - \dfrac{2fm}{rc^2}}\,dr^2 - r^2\,d\Omega^2 \quad (1.4.5)$$

(Birkhoff theorem, Birkhoff, 1923). In this gravitational field, no circular paths obtain for a particle that approaches the gravitational center $r = 0$ to within a distance $r = 3fmc^{-2}$. To compensate the centrifugal force, the particle speed must exceed c for all $r < 3fmc^{-2}$.

According to its linear approximation, general relativity theory asserts that gravitational waves exist. But the physical interpretation of these wave solutions is complicated by the nonlinearity and gauge invariance of general relativity theory: It is not possible to define a localizable energy and a localizable energy flux; physical meaning belongs only to an integral energy flux.

General relativity theory associates the general gravitational field with a breakdown of the Lorentz–Poincaré symmetry of special relativity theory, in the sense that there is no Minkowski background space that allows real active displacements ("motions") of physical systems. In general relativity theory all equations are covariant with respect to transformations of the Einstein group, but this group describes only changes of the coordinate systems (passive displacements) and not real possibilities of displacements of physical systems in space and time. The Lorentz–Poincaré group of special relativity theory is the ten-dimensional Lie group or maximum mobility in space and time (for vanishing curvature;

the de Sitter group replaces the Lie group for constant curvature). In the general theory of gravitation, the gravitational field restricts this mobility and thus the geometric symmetry of space-time; in infinitesimal regions, the mobility is restricted to motions with infinitesimal generators that are solutions to the *Killing equations*: The most general Riemannian space, i.e., the most general gravitational field, does not allow any motions. The genuine integral conservation laws existing in general relativity theory correspond to solutions of the Killing equation

$$-g_{ik,l} \, \xi^l_{K} = g_{il} \, \xi^l_{K,k} + g_{kl} \, \xi^l_{K,i}$$

$$(1.4.6)$$

$$0 \leq K \leq 10$$

(In Rosen's interpretation of the Einstein equations, there exists the Minkowski background space for the g_{ik} field. Here, therefore, the Lorentz–Poincaré group is a true symmetry group and there are ten integral conservation laws.)

If one starts by considering the generalizations of the special relativity principle that arise from the homogeneous part of equation (1.1.1), then the general principle of relativity conveys the invariance of physical equations with respect to local Lorentz transformations. Since the representation space of the local Lorentz group is given by reference tetrads h^A_{i}, this starting point leads to a description of gravitation by means of the reference tetrad fields $h^A_{i} = h^A_{i}(x^l)$. (For the basic concepts of the tetrad formalism and tetrad theories of gravitation, consult Appendix B.) Here the $h^A_{i}(x^l)$ represent the 4×4 components of the space-time reference basis (four orthonormalized standards, one of which is a standard clock, at every world-point). These reference tetrads connect a

Local Principles and the Theory of Gravitation

local proper reference system with the gravitational metric

$$g_{ik} = h^A{}_i h^B{}_k \eta_{AB}$$

$$\eta_{AB} = h^i{}_A h^k{}_B g_{ik}$$

(1.4.7)

An affine non-Lorentzian transformation of the reference tetrads,

$$h^{*A}{}_i = \Omega^A{}_B(x^l) h^B{}_i \tag{1.4.8}$$

generally defines a deformation of the gravitational metric and thus the transition from one gravitational field to another:

$$g^*{}_{ik} = \eta_{CD} \Omega^C{}_A \Omega^D{}_B h^A{}_i h^B{}_k$$

$$= \alpha_{AB}(x^l) h^A{}_i h^B{}_k \tag{1.4.9}$$

For given $g_{ik}(x^l)$, the reference tetrads are determined only up to local Lorentz transformations. (The reference tetrads define the metric spin vectors in a gravitational field by

$$\sigma_i{}^{\alpha\beta} = \sigma_A{}^{\alpha\beta} h^A{}_i$$

where the $\sigma_A{}^{\alpha\beta}$ are the Pauli spin matrices and α and β can each assume the values 1 and 2.)

A description of gravitation by means of the $h^A{}_i$ leads, of course, to a theory of gravitation that does not satisfy the strong equivalence principle, because a tetrad theory requires more than ten functions g_{ik} to characterize the gravitational field. However, the weak equivalence principle, which describes the influence of gravitation on nongravitational matter, provides fewer constraints on the gravitational equations than the strong one; it demands merely that the field equations be compatible with the equations of motion that ensue from the weak equivalence principle.

According to Einstein's equivalence principle, the interaction of the gravitational field with tensorial matter (Bose fields) is determined by the g_{ik} alone. (The effect of gravitation on matter is taken into account through the Einstein-covariant writing of the canonical equations of motion.) The interaction with spinorial matter (Fermi fields), on the other hand, can only be treated with the aid of the $h^A{}_i$. Here the equivalence principle requires the Lorentz-covariant writing of the field equations.

The Einstein-covariant equations for bosons are Lorentz invariant; the Lorentz-invariant field equations for fermions are Einstein invariant. If the $h^A{}_i(x^l)$ were fixed, up to global Lorentz rotations, so that an Einsteinian teleparallelism existed, then only the special relativity principle would remain valid. The following inversion of the conventional interpretation of general and special relativity theory [stressed also by Fock (1960)] is possible: All nongravitational matter satisfies the general principle of relativity. According to the general theory of relativity, this is also true for gravitation. But if gravitation entails a breakdown of symmetries, then gravitation could lead to a definite choice of the $h^A{}_i(x^l)$. In this case, there will exist physical reference systems that are given a privileged status by the gravitational fields; which means that these systems are fixed, up to rigid rotations, by the equations of gravitation plus initial and boundary conditions.

In Galilei–Newton mechanics and in special relativity theory, inertial systems are selected by inertia properties of matter; in general relativity theory they are defined by $h^A{}_i = \delta^A{}_i$. These reference systems are, however, not compatible with the existence of local gravitational fields: Gravitation deforms the $\delta^A{}_i$ into

$h^A{}_i(x^l)$, the deformation, described by $\Omega^A{}_B(x^l)$, being determined by the matter tensor.

The $h^A{}_i$, we know, represent four vector fields. Therefore, in order to formulate fields equations determining the deformations, one has to construct the following 4-vectors out of $h^A{}_i$ and $T_i{}^k$:

$$h^A{}_k T_i{}^k \quad \text{and} \quad h^A{}_i T \qquad (1.4.10)$$

From the requirement that the linear equations of gravitation must result in first approximation from the correct equations, one obtains quite naturally the following system of local equations of gravitation (Treder 1967, 1971b):

$$\Box\, \Omega^A{}_B + \varkappa \Omega^A{}_C \underset{0}{T^{*C}_B} = 0 \qquad (1.4.11)$$

where the asterisked symbols denote the special-relativistic quantities

$$T^{*C}_B \equiv \delta_k{}^C \delta_B{}^i T^{*k}_i \qquad (1.4.12)$$

$$T^{*k}_i \equiv T_i{}^k - \tfrac{1}{2}\delta_i{}^k T_m{}^m$$

Equations (1.4.11) correspond to the following set of equations for the $h^A{}_i$ (Treder, 1967):

$$\Box\, h^A{}_i + \varkappa h^A{}_l T^{*l}_i = 0 \qquad (1.4.13)$$

These equations, being linear and homogeneous with respect to the $h^A{}_i$, are privileged over all other possible equations (Treder, 1967, 1971b). They furnish the complete linear theory—the

Newtonian approximation and Einstein's gravito-optics. With respect to the g_{ik}, equations (1.4.13) are nonlinear (the $h^A{}_i$ are like square roots of the g_{ik}), and correspondingly there exists a post-Newtonian approximation. In particular, this approximation predicts the following value of the perihelion angular motion:

$$\delta\psi = \tfrac{7}{6}(\delta\psi)_{\text{Einstein}} \tag{1.4.14}$$

The decisive difference between equations (1.4.13), on the one hand, and the linear theory and general relativity theory, on the other, is the potentiallike coupling

$$\varkappa T^{*l}{}_i h^A{}_l$$

between matter and the gravitational potential. This coupling gives rise to the general expression

$$m_A = \frac{2}{c^2} \int h^0{}_0 T^{*0}{}_0 \, d^3x \tag{1.4.15}$$

for the active gravitational mass. The inertial mass, by comparison, is given by

$$m_I = \frac{2}{c^2} \int h^1{}_1 T^{*1}{}_1 \, d^3x \tag{1.4.16}$$

Here we have

$$h^0{}_0 < 1$$
$$\tag{1.4.17}$$
$$h^1{}_1 > 1$$

72

from which it follows that

$$m_A < \frac{2}{c^2} \int T^{*0}_{\ 0} \, dV < m_I \qquad (1.4.18)$$

The active gravitational mass decreases with increasing gravitational potential and is always less than the mass computed in general relativity theory and less than the inertial mass, as well. For very large inertial masses ($m_I \to \infty$), the active heavy mass m_A tends to a finite limit, which depends on the mass density and the equations of state. In principle, a volume can contain an arbitrary number of particles, without the exterior gravitational potential of the system exceeding a certain finite limit.

The gravitational action of the majority of particles is screened by the gravitational potential itself. Accordingly, there is no gravitational collapse.

The tetrad field equations are linear with respect to the tetrads $h^A{}_i$; but, as mentioned above, they can be rewritten as nonlinear equations with respect to the metric g_{ik}, which is constructed from the $h^A{}_i$ in accordance with equations (1.4.7). Equations (1.4.13) thus describe genuine nonlinear gravitational effects.

By adding a bilinear term to the left-hand side of the field equations and choosing its coefficient properly, one can assign any desired value to the perihelion advance.

All theories of gravitation have cosmological consequences. There exists, however, at present no possibility to experimentally test for such consequences. *Weyl's cosmological principle*, which is a generalization of Copernicus' world postulate of a homogeneous and isotropic cosmos, asserts that the average state of motion of matter in our universe can be described by a global time-independent rest system with the metric

$$ds^2 = c^2\,dt^2 - R^2(t)\,\frac{d\mathbf{r}^2}{(1 + \varepsilon r^2/4)^2} \qquad (1.4.19)$$

By the cosmological principle, the structure of the cosmos is fixed to a large extent; theories of gravitation play a role only in that they propose different forms of the function $R(t)$.

Chapter 2

Intermezzo: The Einstein Effects

2.1. Significance of the Post-Newtonian Kepler Problem

According to the weak principle of equivalence, the Lagrangian $\mathcal{L}$ describing the motion of a test particle with rest mass m in the gravitational field of a central mass M is given by the relativistic line element ds:

$$\left(c - \frac{\mathcal{L}}{mc}\right) dt = ds = (g_{ik}\, dx^i\, dx^k)^{1/2} \qquad (2.1.1)$$

From Section 1.3, the static, spherically symmetric gravitational potential g_{ik} of the mass M, expressed in isotropic coordinates, for relativistic gravitation theories reads as follows:

$$g_{00} = 1 - 2\frac{fM}{c^2 r} + \chi \frac{f^2 M^2}{c^4 r^2}$$

$$g_{\mu 0} = 0 \qquad (2.1.2)$$

$$g_{\mu\nu} = -\delta_{\mu\nu}\left(1 + 2\omega\,\frac{fM}{c^2 r}\right)$$

where χ and ω are pure numbers. This result is valid in the post-Newtonian approximation for the equations of motion. Equation (2.1.1), in combination with equation (2.1.2), states that

$$1 - \frac{\mathscr{L}}{mc^2} = \frac{1}{c}\left(1 - 2\frac{fM}{c^2r} + \chi\,\frac{f^2M^2}{c^4r^2} - \frac{v^2}{c^2} - 2\omega\,\frac{fM}{c^2r}\,\frac{v^2}{c^2}\right)^{1/2}$$

(2.1.3)

for a test particle with 3-velocity $v^\nu = dx^\nu/dt$. Expansion of equation (2.1.3) yields, in the same post-Newtonian approximation,

$$-\frac{\mathscr{L}}{mc^2} = -\frac{fM}{c^2r} - \frac{v^2}{2c^2} - \frac{v^4}{8c^4} - \left(\omega + \frac{1}{2}\right)\frac{fM}{r}\,\frac{v^2}{c^2}$$

$$+ \frac{1}{2}(\chi - 1)\,\frac{f^2M^2}{c^4r^2}$$

(2.1.4)

This equation defines a classical Lagrangian $\mathscr{L}$ with a post-Newtonian interaction:

$$\mathscr{L} = m\left(\frac{v^2}{2} + \varepsilon\,\frac{v^4}{c^2} + \frac{fM}{r} + \beta\,\frac{fM}{r}\,\frac{v^2}{c^2} + \gamma\,\frac{f^2M^2}{c^2r^2}\right)$$

(2.1.5)

wherein the numbers are furnished, according to equation (2.1.4), by

$$\varepsilon = \tfrac{1}{8}$$

$$\beta = \tfrac{1}{2} + \omega$$

(2.1.6)

$$\gamma = \tfrac{1}{2} - \tfrac{1}{2}\chi$$

With the choice $\chi = 2$ and $\omega = 1$, the metric tensor (2.1.2) becomes the first post-Newtonian approximation of the metric for the Schwarzschild line element

$$ds^2 = -\left(1 + \frac{fM}{2c^2r}\right)^4 \delta_{\mu\nu}\, dx^\mu\, dx^\nu + \frac{(1 - fM/2c^2r)^2}{(1 + fM/2c^2r)^2}\, c^2\, dt^2 \quad (2.1.7)$$

Thus, if one takes

$$\varepsilon = \tfrac{1}{8}$$

$$\beta = \tfrac{3}{2}$$

$$\gamma = -\tfrac{1}{2}$$

then the classical Lagrangian (2.1.5) is dynamically equivalent to the Schwarzschild metric in the post-Newtonian approximation.

The post-Newtonian perihelion precession of the planets is quite generally given by the formula

$$\delta\psi = 2\pi\Omega\, \frac{\alpha}{(1 - e^2)a}$$

$$= 2\pi(4\varepsilon + 2\beta + \gamma)\, \frac{\alpha}{(1 - e^2)a} \quad (2.1.8)$$

(a = major semiaxis, e = eccentricity), $\alpha = fM/c^2$ being the Schwarzschild constant, which was put forward by Tisserand (1872) and Einstein (1915). The interaction potential that gives rise to this result is contained in the Lagrangian

$$\mathscr{L} = m\left(\frac{v^2}{2} + \varepsilon\, \frac{v^4}{c^2} + \alpha\, \frac{c^2}{r} + \beta\, \frac{\alpha}{r}\, v^2 + \gamma\, \frac{\alpha^2 c^2}{r^2}\right)$$

$$= m\left(\frac{v^2}{2} + \varepsilon\, \frac{v^4}{c^2}\right) + m\phi^* \quad (2.1.9)$$

where

$$\phi^* = \frac{\alpha c^2}{r} + \beta \frac{\alpha}{r} v^2 + \gamma \frac{\alpha^2 c^2}{r^2} \qquad (2.1.10)$$

is the effective interaction potential of the heavy mass $M = c^2 \alpha f^{-1}$, which describes action at a distance without retardation effects.

When retarded effects of gravitation are taken into account, ϕ^* in the expressions (2.1.9) and (2.1.10) has to be supplemented by the Gauss–Neumann retardation of the post-Newtonian approximation,

$$\phi^{*\prime} - \phi^* = \delta \frac{\alpha}{r} \dot{r}^2, \qquad \delta \geq 0$$

$$\qquad (2.1.11)$$

$$\dot{r} = \frac{dr}{dt} = \frac{x^\mu v^\mu}{r}$$

Here $\phi^{*\prime}$ corresponds to a retardation of gravitational effects with the speed of propagation

$$c^* = \frac{c}{\delta^{1/2}} \qquad (2.1.12)$$

Following Neumann (1868, 1896), we can derive the foregoing from a postulate advanced by Gauss: Replace the Newtonian potential

$$\phi = -\frac{\alpha c^2}{r}$$

by the "retarded interaction potential"

$$-\phi' = \frac{\alpha c^2}{r(1 - \dot{r}/c^*)} \qquad (2.1.13a)$$

and expand it in terms of the parameter $\dot{r}/c^* \ll 1$:

$$-\phi' \approx \frac{\alpha c^2}{r} \left(1 + \frac{\dot{r}}{c^*} + \frac{\dot{r}^2}{c^{*2}} \right) \qquad (2.1.13\text{b})$$

In view of the equality

$$\frac{\dot{r}}{r} = \frac{d}{dt} (\ln r)$$

it is then seen that the term

$$\frac{\dot{r}}{r} \, \alpha \, \frac{c^2}{c^*}$$

does not contribute to the Lagrangian equations of motion, and one is led to the interaction potential given by equation (2.1.11).

These arguments also involve the objections of Gauss and Poincaré to Laplace's (1798–1825) statement that a gravitational retardation is incompatible with celestial mechanics.

The expression

$$\phi' = \phi_{\text{Weber}} = -\frac{\alpha}{r} \left(c^2 - \dot{r}^2 \frac{c^2}{c^{*2}} \right)$$

$$= \phi - \phi\delta \frac{\dot{r}^2}{c^{*2}} \qquad (2.1.14)$$

is Weber's generalization (1846) of Newton's potential, and

$$\phi_{\text{Riemann}} = -\frac{\alpha c^2}{r} \left(1 - \beta \frac{v^2}{c^2} \right)$$

$$= \phi - \phi\beta \frac{v^2}{c^2} \qquad (2.1.15)$$

is Riemann's isotropic generalization of Newton's potential (1858); cf. Treder (1972, 1975).

The general Lagrangian with retardation,

$$\mathscr{L}' = m\left(\frac{v^2}{2} + \varepsilon\,\frac{v^4}{c^2}\right) + m\phi^* + m\,\frac{\alpha}{r}\,\delta\dot{r}^2 \qquad (2.1.16)$$

yields the constant

$$\Omega' = \Omega + \delta$$

$$= 4\varepsilon + 2\beta + \gamma + \delta \qquad (2.1.17)$$

for the perihelion advance.

In classical theories of gravitation, proposed by Riemann (1875), Neumann (1868, 1896), Zöllner (1872), Tisserand (1895, 1896), Gerber (1898, 1917), and others (see Oppenheim, 1920), one always has $\varepsilon = 0$ because this is a necessary condition for the Galilean invariance of the Lagrangian $\mathscr{L}$. If no direction of motion is preferred, then the Lagrangian must also be isotropic. In this case one has $\delta = 0$ and accordingly the general post-Newtonian classical gravitational dynamics follows from the Lagrangian

$$\mathscr{L}_{\text{classical}} = m\left(\frac{v^2}{2} + \frac{\alpha c^2}{r} + \beta\,\frac{\alpha}{r}\,v^2 + \gamma\,\frac{\alpha^2 c^2}{r^2}\right)$$

$$\Omega = 2\beta + \gamma \qquad (2.1.18)$$

On the other hand, in Lorentz-covariant relativistic theories one always has $\varepsilon = \frac{1}{8}$, which yields

$$\mathscr{L}_{\text{relativistic}} = m\left(\frac{v^2}{2} + \frac{v^4}{8c^2} + \frac{\alpha c^2}{r} + \beta\,\frac{\alpha}{r}\,v^2 + \gamma\,\frac{\alpha^2 c^2}{r^2}\right) \qquad (2.1.19)$$

Intermezzo: The Einstein Effects

under isotropic circumstances ($\delta = 0$), with

$$\Omega_{\text{relativistic}} = \tfrac{1}{2} + 2\beta + \gamma \qquad (2.1.20)$$

A retardation is always equivalent to anisotropy of the effective interaction potential

$$\phi^{*\prime} = \frac{\alpha c^2}{r} + \frac{\alpha c^2}{r}\left(\beta\frac{v^2}{c^2} + \gamma\frac{\alpha}{r} + \frac{\delta\dot{r}^2}{c^2}\right)$$

$$= \phi^* - \phi\delta\frac{\dot{r}^2}{c^2} \qquad (2.1.21)$$

When the effective speed of propagation

$$c^* = \frac{c}{\delta^{1/2}}$$

decreases, this anisotropy increases in proportion to c^{*-2}. [Gerber's potential is the Neumann potential (2.1.21) with $\delta = 3$.]

On replacing the retarded interaction potential (2.1.6),

$$-\phi' = \frac{fM}{r(1 - \dot{r}/c^*)}$$

$$\approx \frac{fM}{r}\left(1 + \frac{\dot{r}}{c^*} + \frac{\dot{r}^2}{c^{*2}}\right) \qquad (2.1.22)$$

by an advanced potential in the post-Newtonian limit,

$$-\phi'' = \frac{fM}{r(1 + \dot{r}/c^*)}$$

$$\approx \frac{fM}{r}\left(1 - \frac{\dot{r}}{c^*} + \frac{\dot{r}^2}{c^{*2}}\right) \qquad (2.1.23)$$

we obtain the Neumann–Weber interaction potential

$$\phi^{*\prime\prime} = \phi^{*\prime} = \phi^* + \frac{\alpha}{r}\,\frac{c^2}{c^{*2}}\,\dot{r}^2$$

$$= \phi^* + \delta\,\frac{\alpha}{r}\,\dot{r}^2 \qquad (2.1.24)$$

Because of the above quadratic dependence of the post-Newtonian interaction potentials $\phi^{*\prime}$ and $\phi^{*\prime\prime}$ on the propagation speed of gravitation, the effective Lagrangian $\mathscr{L}'$, corresponding to the potentials ϕ' and ϕ'', respectively, is time-symmetric. Therefore, $\mathscr{L}'$ can also be regarded as a Lagrangian of a stationary (Weber–Riemann) interaction potential. If the Weber term

$$\frac{1}{m}\,(\mathscr{L}' - \mathscr{L}) = \phi^{*\prime} - \phi^*$$

$$= -\delta\,\frac{\alpha}{r}\,\dot{r}^2$$

is negative, i.e., $\delta > 0$, this corresponds to an imaginary speed of retardation

$$c^* = \frac{c}{(-\delta)^{1/2}}$$

Such a negative term can therefore not be interpreted as the result of a finite gravitational speed of propagation.

The possibility for expressing the retarded interaction by a time-independent effective interaction potential [here the potential (2.1.21)] exists not only in the post-Newtonian approximation of celestial mechanics. Indeed, in the corresponding $v^2 c^{-2}$ approximation of electrodynamics ("post-Coulomb approximation"), the

Liénard–Wiechert potential can be transformed into the Helmholtz–Darwin interaction potential, which is also time-independent and bilinear in the speeds of the carriers of charge.

In his general theory of relativity, Einstein (1915, 1916) derived the formula (2.1.1) in a completely novel manner from the geodesic law of motion. One has

$$\frac{D}{D\tau} P_l = 0 \leftrightarrow m\left(\frac{d}{d\tau} (g_{lk}u^k) - \frac{1}{2} \partial_l g_{mn} u^m u^n \right) = 0 \quad (2.1.25)$$

with

$$P_l = mg_{lk}u^k$$

$$u^k = \frac{dx^k}{d\tau} \quad (2.1.26)$$

$$c^2 \, d\tau^2 = g_{mn} \, dx^m \, dx^n$$

On substituting in equation (2.1.25) the formulas for the second-order terms of the spherically symmetric metric belonging to a point mass

$$M = \frac{\alpha c^2}{f}$$

one gets

$$g_{rr} = -\left(1 + 2\omega \frac{\alpha}{r} + 2\lambda \frac{\alpha}{r}\right)$$

$$g_{\varphi\varphi} = -r^2\left(1 + 2\omega \frac{\alpha}{r}\right) \quad (2.1.27)$$

$$g_{00} = 1 - 2\tau \frac{\alpha}{r} + \chi \frac{\alpha^2}{r^2}$$

Einstein (1915) obtained the general formula

$$\delta\psi = 2\pi\Omega' \, \frac{\alpha}{a(1 - e^2)}$$

$$\Omega' = 2\omega + 2\tau - \frac{\chi}{2\tau} + \lambda$$

(2.1.28)

for the perihelion shift. In particular, employing Gaussian coordinates, one finds for the Schwarzschild metric in general relativity theory (Schwarzschild, 1916) that

$$\omega_S = 0$$

$$\chi_S = 0$$

$$\lambda_S = 1$$

$$\tau_S = 1$$

(2.1.29)

whereas use of Weyl's isotropic coordinates [see equation (2.1.2)], leads to

$$\omega_W = 1$$

$$\chi_W = 2$$

$$\lambda_W = 0$$

$$\tau_W = 1$$

(2.1.30)

It follows that

$$\Omega' = \Omega_S = 2\tau_S + \lambda_S = 2 + 1$$

$$= \Omega_W = 2\omega_W + 2\tau_W - \frac{\chi_W}{2\tau_W} = 2 + 2 - 1 \qquad (2.1.31)$$

Intermezzo: The Einstein Effects

The connection between the values (2.1.29) and (2.1.30) is provided by the transformation

$$r_S = r_W\left(1 + \frac{\alpha}{2r_W}\right)^2$$

$$\approx r_W + \alpha \tag{2.1.32}$$

In a coordinate system where $g_{\alpha\beta}$ is isotropic, the spherically symmetric metric coefficients generally are

$$g_{\mu\nu} = -\delta_{\mu\nu}\left(1 + 2\omega\frac{\alpha}{r}\right)$$

$$g_{00} = 1 - 2\tau\frac{\alpha}{r} + \chi\frac{\alpha^2}{r^2} \tag{2.1.33}$$

and consequently $\lambda = 0$. This leads to

$$\Omega' = \Omega$$

$$= 2\omega + 2\tau - \frac{\chi}{2\tau} \tag{2.1.34}$$

In the first-order limit, the Newtonian law of motion,

$$-\dot{v}^i = \frac{\partial}{\partial x^i}g_{00}\frac{c^2}{2}$$

$$= \tau\frac{\partial}{\partial x^i}\phi$$

$$= \frac{fM}{r^3}x^i \tag{2.1.35}$$

must be valid, whence $\tau = 1$. One thus obtains the following relativistic factor in the Tisserand–Einstein formula:

$$\Omega = 2 + 2\omega - \tfrac{1}{2}\chi$$

$$\Omega' = 2 + 2\omega - \tfrac{1}{2}\chi + \lambda$$

(2.1.36)

Finally, a term-by-term comparison of Einstein's geodesic law of motion with the post-Newtonian Lagrange equations of motion gives

$$\beta = \omega + \tfrac{1}{2}$$

$$\gamma = \tfrac{1}{2} - \tfrac{1}{2}\chi$$

$$\Omega = \tfrac{1}{2} + 2\beta + \gamma$$

$$\Omega' = \Omega + \lambda$$

(2.1.37)

[Levi-Civita (1947); Einstein, Infeld, and Hoffmann (1938; 1940); Fock (1955)], from which the identity of Einstein's and Tisserand's formulas ensues. With the aid of (2.1.37), equation (2.1.16) furnishes the interaction potential of general relativity theory for the post-Newtonian approximation of the planetary motion.

On applying the transformation equation

$$\bar{r} = r + \lambda\alpha$$

(2.1.38)

for the radial coordinate to the isotropic form (2.1.33) of a spherically symmetric space–time metric in the post-Newtonian limit, one finds

$$g_{\bar{r}\bar{r}} = -\left(1 + 2\omega\,\frac{\alpha}{\bar{r} - \lambda\alpha}\right)$$

$$\approx -\left(1 + 2\omega\,\frac{\alpha}{\bar{r}}\right) \tag{2.1.39a}$$

$$g_{\psi\psi} = -(\bar{r} - \lambda\alpha)^2\left(1 + 2\omega\,\frac{\alpha}{\bar{r} - \lambda\alpha}\right)$$

$$\approx -\bar{r}^2\left(1 - \frac{\lambda\alpha}{\bar{r}}\right)^2\left(1 + 2\omega\,\frac{\alpha}{\bar{r}}\right)$$

$$\approx -\bar{r}^2\left[1 + 2(\omega - \lambda)\,\frac{\alpha}{\bar{r}}\right] \tag{2.1.39b}$$

$$g_{00} = 1 - \frac{2\alpha}{\bar{r} - \lambda\alpha} + \chi\,\frac{\alpha^2}{(\bar{r} - \lambda\alpha)^2}$$

$$\approx 1 - \frac{2\alpha}{\bar{r}} + (\chi - 2\lambda)\,\frac{\alpha^2}{\bar{r}^2} \tag{2.1.39c}$$

Thus, if one uses the new constants,

$$\bar{\omega} = \omega - \lambda$$

$$\bar{\chi} = \chi - 2\lambda$$

the general anisotropic form (2.1.27) is recovered:

$$-g_{\bar{r}\bar{r}} = 1 + 2(\bar{\omega} + \lambda)\,\frac{\alpha}{\bar{r}} \tag{2.1.40a}$$

$$-g_{\psi\psi} = \bar{r}^2\left(1 + 2\bar{\omega}\,\frac{\alpha}{\bar{r}}\right) \tag{2.1.40b}$$

$$g_{00} = 1 - \frac{2\alpha}{\bar{r}} + \bar{\chi}\,\frac{\alpha^2}{\bar{r}^2} \tag{2.1.40c}$$

The transformed Lagrangian (2.1.19) reads

$$\mathscr{L} = m\left(\frac{v^2}{2} + \frac{1}{8}\frac{v^4}{c^2}\right) + m\left(\frac{\alpha c^2}{\bar{r}} + \bar{\beta}\frac{\alpha}{\bar{r}}v^2 + \bar{\gamma}\frac{\alpha^2 c^2}{\bar{r}^2} + \delta\frac{\alpha}{\bar{r}}\dot{r}^2\right)$$

$$(2.1.41)$$

with the constants

$$\bar{\beta} = \bar{\omega} + \tfrac{1}{2} = \omega - \lambda + \tfrac{1}{2} = \beta - \lambda$$

$$\bar{\gamma} = \tfrac{1}{2} - \tfrac{1}{2}\bar{\chi} = \tfrac{1}{2} - \tfrac{1}{2}\chi + \lambda = \gamma + \lambda \qquad (2.1.42)$$

$$\delta = \lambda$$

Equation (2.1.41) of course furnishes the same perihelion advance as equation (2.1.19), since one has

$$\bar{\Omega} = \tfrac{1}{2} + 2\bar{\beta} + \bar{\gamma} + \delta$$

$$= \tfrac{1}{2} + 2\beta - 2\lambda + \gamma + \lambda + \lambda$$

$$= \Omega \qquad (2.1.43)$$

We may split the Lagrangian (2.1.41) into an isotropic part $\mathscr{L}^0$ and an anisotropic part $\mathscr{L}'$,

$$\mathscr{L} = \mathscr{L}^0 + \mathscr{L}'$$

$$= \mathscr{L}^0 + m\lambda\frac{\alpha}{\bar{r}}\dot{r}^2 \qquad (2.1.44)$$

Then it is seen, from (2.1.11), that the contribution

$$\frac{\mathscr{L}'}{m} = \lambda\frac{\alpha}{\bar{r}}\dot{r}^2$$

$$\approx \lambda\frac{\alpha}{r}\dot{r}^2 \qquad (2.1.45)$$

Intermezzo: The Einstein Effects

is Neumann's retardation term in the interaction potential. There-fore, with the aid of the metric (2.1.40) and the Lagrangian (2.1.41), the gravitational interaction is found to propagate at the speed given by equation (2.1.12),

$$c^* = \frac{c}{\lambda^{1/2}} \qquad (2.1.46)$$

This means that, in order to interpret the metric (2.1.39) as a representation of retarded gravitation, one has to set $\lambda > 0$ in the transformation formula (2.1.38).

The relation between the isotropic Weyl form of the spherically symmetric line element in general relativity theory and its original Gaussian form [used by Einstein (1915) in his first calculation of the perihelion advance] is given by

$$\bar{r} = \left(1 + \frac{\alpha}{2r}\right)^2 r \qquad (2.1.47)$$

so that, in the post-Newtonian limit,

$$\bar{r} \approx r + \alpha$$
$$\lambda = 1 \qquad (2.1.48)$$

According to equations (2.1.35) and (2.1.36), this corresponds to a gravitational retardation with a speed of propagation $c^* = c$. Thus, in the post-Newtonian limit, the Einstein–Schwarzschild form of the spherically symmetric coefficients g_{ik} implies in general relativity theory a (Neumann) retardation of gravitation (2.1.13), that is, the replacement of Newton's gravitational potential by that of Weber, corresponding to a propagation of gravitation with the speed of light c. On the other hand, Weyl's isotropic form of the spherically symmetric metric [of Fock's harmonic form or, generally, the representation of this metric in accordance with the Einstein–Infeld–Hoffmann conditions] implies an action-at-a-distance representation of the gravitational interaction ($\lambda = 0$, $c^* \to \infty$). Therefore, in the post-Newtonian limit of general relativity theory, the difference between the local and the action-at-a-distance representations becomes a question of the choice of coordinate system.

If the Lagrangian (2.1.9) is supplemented by the additional retardation term

$$m\delta \frac{\alpha}{r} \dot{r}^2$$

and the coordinate r is subjected to the transformation (2.1.38), then one is led to the Lagrangian

$$\mathscr{L}^{*\prime} = m\left(\frac{v^2}{2} + \varepsilon \frac{v^4}{c^2} \right) + m\left(\frac{\alpha c^2}{r^*} + (\gamma + \lambda) \frac{\alpha^2 c^2}{r^{*2}} + \beta \frac{\alpha}{r^*} v^2 \right)$$

$$+ m\delta \frac{\alpha}{r^*} \dot{r}^2 \tag{2.1.49}$$

with the Tisserand–Einstein constant

$$\Omega^{*\prime} = 4\varepsilon + 2\beta + \gamma + \lambda + \delta$$

$$= \Omega + \lambda + \delta \tag{2.1.50}$$

Consequently, a retardation with the speed $c^* = c/\delta^{1/2}$ can be compensated by the transformation $r^* = r - \lambda\alpha$ of the radial coordinate.

Conversely, a retardation of the post-Newtonian gravitational effects with speed c^* as above is dynamically equivalent to a replacement of r by $r^* = r - \delta\alpha$ in the Lagrangian. Indeed, one finds

$$\mathscr{L}^* = m\left(\frac{v^2}{2} + \varepsilon\,\frac{v^4}{c^2}\right) + m\left(\frac{\alpha c^2}{r^*} + \beta\,\frac{\alpha}{r^*}\,v^2 + \gamma\,\frac{\alpha^2 c^2}{r^{*2}}\right)$$

$$\approx m\left(\frac{v^2}{2} + \varepsilon\,\frac{v^4}{c^2}\right) + m\left(\frac{\alpha c^2}{r} + \beta\,\frac{\alpha}{r}\,v^2 + (\gamma + \delta)\,\frac{\alpha^2 c^2}{r^2}\right)$$

$$(2.1.51)$$

and from this it follows again that

$$\Omega^* = \Omega + \delta$$

$$= \Omega'$$

The general-relativistic theory of gravitodynamics can there-fore be formulated in a form that is free of retardation effects in the post-Newtonian limit. When applied to the motion of a single body in a spherically symmetric gravitational field g_{ik}, this representation corresponds to the choice of an isotropic form $g_{\mu\nu} \sim -\delta_{\mu\nu}$ for the spatial metric, which in turn leads to an isotropic Lagrangian $\mathscr{L}$.

If the *background metric* (the *world background*, after Wie-chert), relative to which the interaction potential is defined, is measured with the help of rigid measuring rods and thus defined in the Euclidean sense, then the retardation of the gravitational interaction has a physical meaning. With respect to a world back-ground that is measured by rigid bodies serving as standards, the local and the action-at-a-distance viewpoints give rise to different physical consequences in the post-Newtonian (classical and rela-tivistic) theories of gravitation. Under these conditions, the effective retardation speed of gravitation $c^* = c/\delta^{1/2}$ is well defined. By contrast, if, in accordance with Einstein's covariance principle, all Gaussian coordinate systems are viewed as equivalent (i.e., the world background is measured by means of Einstein's "Bezugs-mollusken"), then retardation in the post-Newtonian limit is merely a consequence of the choice of Gaussian coordinates: Fixing the retardation speed c^* is equivalent to choosing a particular Gaussian coordinate system. Accordingly, the general covariance of the interaction potential, i.e., of the post-Newtonian Lagrangian, simply means that all physical post-Newtonian effects are independent of the retardation velocity.

On the other hand, if one refers to a Euclidean (or Minkowski) background metric, in the sense of a bimetric interpretation of relativistic theories of gravitation, and thereby attributes an absolute meaning to the Gaussian coordinates (i.e., to the coordinate conditions), one may decide on Weyl's isotropic form (2.1.33) or on the Einstein–Schwarzschild–Gauss form (2.1.32) of the spherically symmetric gravitational field. Then the isotropic form can be regarded as a representation of an instantaneous action-at-a-distance conception of gravitation with $c^* \rightarrow \infty$ (and without an "ether"). Accordingly, in this case the Gaussian form would correspond to the local conception in the sense of a theory of special relativity (with a Lorentz ether), where gravitational effects propagate at light speed, $c^* = c$.

As mentioned above, it was pointed out in the previous century (Isenkrahe, 1879) that the hypothesis of a finite retardation velocity of gravitation is related to the idea of a medium acting as a carrier of gravitation. Then, as mentioned by Mach (1912) in his analysis of some classical post-Newtonian theories of gravitation (in connection with his remarks on retarded gravitational potentials), the speed $c^* = c$ of gravitation suggests that the ether of light and electromagnetism is also the "medium of gravity," i.e., a consequence of the "unity of forces." This was also the view taken by Faraday and Lorentz and, later, by Mie, Hilbert, Weyl,

Intermezzo: The Einstein Effects

Eddington, and Einstein in their unified field theories (see Section 2.2).

One might be surprised by the fact that, in the post-Newtonian approximation, i.e., to terms of the order $(\alpha c^2/r)(v^2/c^2)$, the Schwarzschild metric in anisotropic coordinates, for which it is explicitly time-independent, makes provision for a retardation of gravitation. It should therefore be pointed out that Weber's interaction potential

$$-\phi' = \frac{\alpha}{r}\,(c^2 + \delta\dot{r}^2) \qquad (2.1.52)$$

is not only the result of an expansion of the retarded gravitational potential

$$\phi = \frac{-\alpha c^2}{r(1 - \dot{r}/c^*)} \qquad (2.1.53)$$

with

$$c^{*2} = \frac{c^2}{\delta}$$

to terms of order v^2/c^2 and, accordingly, an approximate solution to the wave equation

$$\left(\varDelta - \delta\,\frac{\partial^2}{\partial t^2}\right)\phi' = 0 \qquad (2.1.54)$$

it is also an exact solution of the biharmonic potential equation

$$-\varDelta\varDelta\phi' = 2\alpha\delta v^\mu v^\nu\,\varDelta\!\left(\frac{\delta_{\mu\nu}}{r^3} - \frac{3x_\mu x_\nu}{r^5}\right)$$
$$= 0 \qquad (2.1.55)$$

93

see Riemann (1875), Treder (1975). Moreover, in view of well-known facts from electrodynamics, it should not come as a surprise that the Helmholtz–Darwin potential $A_k = (A_0, A_r)$ of electrodynamics can be regarded as the v^2/c^2 limit deriving from the Liénard–Wiechert (retarded) solution of the wave equation. At the same time, the potentials A_0 and A_i are also exact solutions of modified Poisson equations.

In Einstein's general relativity theory, a retardation of the gravitational action of a spherically symmetric mass $M = \alpha c^2/f$ is a pure coordinate effect. For, in general relativity theory, according to Birkhoff's theorem, the (essentially time-independent) Schwarzschild metric is already the general spherically symmetric solution of Einstein's vacuum equations $R_{ik} = 0$. On the other hand, in general-covariant theories of gravitation differing from Einstein's general relativity theory (and generalizing it), gravitational retardation in the post-Newtonian solution of the Kepler problem is not caused by the choice of coordinates, since these relativistic theories of gravitation generally possess essentially time-dependent spherically symmetric solutions of their field equations.

Referring to equation (2.1.10), we note that use of the effective potential ϕ^* is equivalent to the introduction of a classical Lagrangian containing a dipolelike correction term to the Newtonian potential ϕ.

We get, in the latter case,

$$\mathscr{L}^{**} = \frac{m}{2}\, v^2 + m\phi^{**}$$

$$= \frac{m}{2}\, v^2 + m\left(\frac{fM^*}{r} + \Omega\, \frac{M^2 f^2}{c^2 r^2}\right) \qquad (2.1.56)$$

wherein $-\Omega M^2 f^2/c^2 r^2$ is the dipolelike correction to ϕ as studied

Intermezzo: The Einstein Effects

by Einstein and Infeld (1949). The characteristic of the Tisserand–Einstein formula (2.1.8), which derives from this modified Lagrangian, is its prediction of a reciprocal proportionality of the perihelion advance $\delta\psi$ to the semimajor axis a of the planets.

On the other hand, if we consider Newton's law of gravitation and use Laplace's perturbation potential for the classical three-body potential, the derivation of $\delta\psi$ from the secular perturbations of the Kepler motion by a mass ΔM (hidden or as yet unknown) leads to a quite different dependence of $\delta\psi$ on a. If—in accordance with Leverrier's hypothesis on the existence of a planet "Vulcanus," in his "Théorie et tables du mouvement de Mercure" (1859)—one interprets ΔM as the mass of a planet with orbit lying inside that of Mercury, then Leverrier's formula,

$$\delta\psi = \frac{3}{2}\,\pi\,\frac{\Delta M \varrho^2}{M x^2} \qquad (\varrho < a) \qquad (2.1.57)$$

is obtained [see Chazy (1928, 1930)]; ϱ denotes the semiaxis of the Vulcanus orbit. The same formula can also be derived from the potential of a quadrupolelike distribution of mass. Accordingly, equation (2.1.57) is to be viewed as the result of a quadrupolelike, rather than a dipolelike, correction to the Newtonian potential.

Following Gauss' theory of secular perturbations, one may treat the mass ΔM of a planet, in agreement with Kepler's second law, as being distributed along the planetary orbit. Accordingly, Leverrier's submercurial planet may be regarded as an inner ring of planetoids (Harzer 1891, 1896). If it is now assumed that $\varrho = R$, R being the solar radius, the mass ΔM will become part of the sun. This implies a "swelling" that is equivalent to an oblateness or a gravitational quadrupole moment of the sun (see New-

95

comb, 1895), and which gives rise to the law

$$\delta\psi = \frac{6}{5} \, \pi \, \frac{\Delta R}{R} \, \frac{R^2}{a^2} \tag{2.1.58}$$

for the perihelion shift. (This formula was popularized again by Dicke's discussions of the Einstein effects in the scalar–tensor theories of gravitation.) According to von Seeliger (1906), there possibly exist dustlike distributions of an additional mass ΔM in the solar system, such that the corresponding integral secular perturbations (in the sense of Gauss) do provide Tisserand's expression for the perihelion shift. The matter distribution of the zodiacal light assumed by von Seeliger constitutes just such a distribution. In the pure planetary theory it is impossible to distinguish between von Seeliger's hidden masses with universal gravitation and the relativistic and nonrelativistic post-Newtonian corrections to the law of gravitation. (It must be mentioned, however, that von Seeliger's hypothesis requires in general that the Gaussian gravitational constant of the sun system not be defined in the same manner as in post-Newtonian theories; see below.)

Empirically, the Tisserand–Einstein formula with the weight factor $\Omega \approx 3$ is well confirmed. In celestial mechanics there is hardly any room left for a perturbation term of the Leverrier type, i.e., for a planet Vulcanus or an inner ring of planets. The combined effect of post-Newtonian gravitational effects and the mass ΔM of Vulcanus would give rise to a perihelion advance

$$\delta\psi = 2\pi\Omega \, \frac{fMc^{-2}}{a(1 - e^2)} + \frac{3}{2} \, \pi \, \frac{\Delta M \varrho^2}{Ma^2} \tag{2.1.59}$$

However, the (Newtonian) empirical second term can amount to only a few per cent of the non-Newtonian first term.

Intermezzo: The Einstein Effects

Some other modifications of Newton's law exist which predict perihelion advances that are different from that of the Tisserand–Einstein formula (2.1.8). But these modifications involve also constants other than a and fMc^{-2}. Nevertheless, such non-Newtonian laws of gravitational forces

$$\mathfrak{F}_r = -\frac{\partial \phi}{\partial r}$$

are studied in some early papers on non-Newtonian effects in celestial mechanics [Laplace (1798–1825), Tisserand (1895, 1896), Chazy (1928, 1930), Oppenheim (1920)].

The gravitational forces that were proposed by Clairaut and by Euler,

$$-\frac{\partial \phi^+}{\partial r} = \frac{fM}{r^2} + A \frac{fM}{r^{2+\beta}}$$

$$= \mathfrak{F}_r^+ \tag{2.1.60}$$

with $\beta > 0$ and $|A| \ll a^\beta$, furnish perihelion advances

$$\delta^+\psi \approx \pi\beta Aa(1 - e^2)^{-\beta} \tag{2.1.61}$$

An expression for $\delta\psi$ that depends neither on a nor on fM results from the law

$$-\frac{\partial \phi^-}{\partial r} = f^* \frac{M}{r^{2+\beta}}$$

$$= \frac{g}{r^\beta} \frac{fM}{r^2}$$

$$= \mathfrak{F}_r^- \tag{2.1.62}$$

with β small. This law was discussed by Hall (1895) and Newcomb (1895); it gives a perihelion advance

$$\delta\psi \approx \beta\pi \tag{2.1.63}$$

However, the first author to quote this formula was Newton himself, in the first edition (1687) of his *Principia Mathematica*.

Because of the interpretations of the new constants A and g, appearing in (2.1.60) and (2.1.62), respectively, only the Tisserand–Einstein expression for $\delta\psi$ is compatible with our postulates presented in Chapter 1. In Chapter 3, we shall prove that gravitational theories that are based on the global (Mach–Einstein) aspects of gravitation also give rise to the Tisserand–Einstein formula. In particular, we shall deduce, from Mach's relativity of inertia principle and the Mach–Einstein doctrine, the purely gravitational origin of inertia described by a Riemannian interaction potential with $\beta = \frac{3}{2}$, which leads to the perihelion advance

$$\delta\psi = 4\pi\beta \, \frac{fM}{c^2 a(1 - e^2)}$$

Some nonrelativistic ($\varepsilon = 0$) combinations of post-Newtonian interaction potentials (2.1.11) with retardation effects were proposed at the time that Einstein founded his general relativity theory. Painlevé, among others, reworked the earlier theories of Tisserand, Zöllner, Neumann, Lévy, and Gerber to obtain classical explanations of Einstein's effects. Of course, in the Kepler problem all these efforts lead only to special cases of the Lagrangian (2.1.16) (with $\varepsilon = 0$) and special expressions for the interaction potential (2.1.11). Some other quite relativistic attempts are based on the Lagrangian (2.1.8) with $\varepsilon = \frac{1}{8}$. Such theories correspond to Wiechert's quasiclassical interpretation of general relativity theory as a theory of a "structured ether." Wiechert (1916) himself proposed a Lagrangian without "retardation" [cf. Wiechert (1925)]:

$$\mathscr{L}_{\text{Wiechert}} = m\left(\frac{v^2}{2} + \frac{v^4}{8c^2}\right) + m\left[\frac{fM}{r} + \left(\Omega - \frac{1}{2}\right)\frac{f^2M^2}{c^2r^2}\right]$$

$$(2.1.64)$$

By the way, also the so-called post-Newtonian approximation theories of gravitation, which are nowadays often discussed, do little more than reproduce the considerations of Painlevé or Wiechert. To all such "generalized" frameworks for gravitodynamics Einstein's criticism applies (Einstein, 1948):

> Bei der engeren Gruppe bleiben und gleichzeitig die komplizierte Struktur der allgemeinen Relativitätstheorie zugrunde zu legen, bedeutet eine naive Inkonsequenz.*

(Here, Einstein's notion of a "general group" refers to the general covariance.)

A theoretical correction to the Einsteinian value $\Omega = 3$ for the perihelion advance results from the combined theory of gravitational and electromagnetic fields, that is, from the Einstein–Maxwell equation of general relativity theory. This correction also occurs in the event that the central mass $M = \alpha c^2/f$ carries a free electric charge Q. Then the metric g_{ik} is given by the Reissner–Nordström solution instead of the Schwarzschild metric. In Gaussian coordinates, the Reissner–Nordström potentials of a mass M with charge

---

* *Translation*: It signifies a naive inconsistency keeping to the narrower group while simultaneously basing general relativity theory upon the complicated structure.

Q are the coefficients of a spherically symmetric line element

$$ds^2 = \left(1 - \frac{2\alpha}{r} + \frac{f}{c^4} \frac{Q^2}{r^2}\right)c^2\, dt^2$$

$$- \left(1 - \frac{2\alpha}{r} + \frac{f}{c^4} \frac{Q^2}{r^2}\right)^{-1} dr^2 - r^2\, d\omega^2 \qquad (2.1.65)$$

and the Coulomb potential

$$A_i = \frac{Q}{r}\, \delta_i^0 \qquad (2.1.66)$$

According to equations (2.1.27) and (2.1.37), the metric of equation (2.1.65) corresponds to the post-Newtonian interaction potential

$$\phi^{*\prime}_{\text{Reissner–Nordström}} = \frac{\alpha c^2}{r} + \frac{1}{2} \frac{\alpha}{r} v^2 + \frac{1}{2} \frac{\alpha^2 c^2}{r^2} + \frac{\alpha}{r} \dot{r}^2$$

$$- \frac{1}{2} \frac{Q^2}{fM^2} \frac{\alpha^2 c^2}{r^2} \qquad (2.1.67)$$

Use of this potential leads to a relativistic perihelion advance with the Einstein–Tisserand constant

$$\Omega'_{\text{Reissner–Nordström}} = \frac{1}{2} + \beta + \gamma + \lambda - \frac{1}{2} \frac{Q^2}{fM^2} = 3 - \frac{1}{2} \frac{Q^2}{fM^2}$$

$$(2.1.68)$$

However, the Einstein–Schwarzschild value of the perihelion advance is very well confirmed for the planets in our solar system. Equation (2.1.68) therefore implies an estimate of the electric charge Q of the sun; it must be such that

$$Q^2 < fM^2, \qquad Q < f^{1/2}M \approx 5 \times 10^{29}\ \text{cgs} \qquad (2.1.69)$$

a numerical restriction that is in good agreement with the findings of solar physics.

[We remark parenthetically that our estimate (2.1.69) is valid only in general relativity theory. Unified field theories may give gravitational metrics different from that of equations (2.1.65) and (2.1.66). When condition (2.1.69) obtains, then the equation $g_{00} = 0$ has real roots with a total Einstein shift. But, for $Q^2 > fM^2$, the roots of $g_{00} = 0$ become imaginary. Einstein and Rosen, however, postulated that the metrics associated with their geometrodynamical particle models must have real roots of g_{00} for every value of the particle charge Q. They therefore further hypothesized that the electric charges are imaginary constants $Q^* = iQ$ in geometro-dynamics and in unified field theories.]

2.2. The Significance of Gravitational Optics

In discussions of the physical principles underlying rival relativistic theories of gravitation, two categories of experiments stand out as being of basic importance for the establishment of gravitational theory.

On the one hand, we are concerned with experiments that enable us to empirically compare the inertial mass m_I, the active heavy mass m_A, and the passive heavy mass m_P with each other and therefore to put to test the various equivalence principles (see Chapter 1, Section 3).

On the other hand, fundamental importance must be attached to gravito-optical experiments dealing with red shift and the deviation and delay of light in gravitational fields. As we mentioned earlier, the results of gravito-optical experiments show that the Newtonian laws cannot be extended to cases where $v \approx c$: While light should be accelerated in external gravitational fields according to Newton's gravitational optics, radioastronomical determinations of light speed establish that light is actually slowed down when acted on by gravitation. And this defect of Newtonian theory one cannot overcome by enlarging the theory through the introduction of velocity-dependent potentials (Treder, 1975). Such potentials were applied also to gravito-optics, after the example provided by post-Newtonian gravitational dynamics (see Section 1), by Painlevé. It turns out, however, that in a thus modified Newtonian gravitational theory not enough disposible parameters are available to reconcile the theory with the results of the three gravito-optical experiments.

The great step forward made by the linear equations of gravitation (1.2.21) lies primarily in the fact that—as we explained, when enumerating the advantages of the theory, under point (1) in Section 1.2—they furnish the correct gravito-optics and that the light deviation and red-shift effects (together with the special relativity principle and the special-relativistic formulation of the equivalence principle) uniquely determine the linearized gravitational equations.

The fundamental significance of these effects resides in their ability to test the scope of the principles lying at the basis of gravitational theory, particularly in reference to the compatibility of these principles with the principles of nongravitational physics—a point that was specially emphasized by Einstein when he first formulated his general relativity theory. The bending of light in a gravitational field appeared to Einstein to be a particularly characteristic effect for his conception of gravitation.

Actually, the coupling of the gravitational field with the electromagnetic field, the only other macroscopic field, is of the greatest importance for the physics of fields, a fact that was also theoretically attested to in the attempts of Weyl, Kaluza, Eddington, Einstein, Schrödinger, and others to construct a unitary theory of gravitation and electromagnetism.

It should be stressed that the gravito-optical effects are more important for the experimental confirmation of gravitational theory than the perihelion advance of celestial mechanics. Indeed, expanding the spherical static solution of gravitational theory (i.e., Einstein's, or some other competing, theory), for $r \gg 2fM/c^2$, in powers of $fM/rc^2 \equiv |\phi|$ $(c = 1)$, one gets, from equation (2.1.4),

$$ds^2 = -[1 + \alpha \, | \, \phi \, | + O(\phi^2)] \, dr^2 - [1 + \beta \, | \, \phi \, | + O(\phi^2)]r^2 \, d\omega^2$$
$$+ [1 - 2 \, | \, \phi \, | + \delta\phi^2 + O(\phi^3)] \, dt^2 \qquad (2.2.1)$$

And if this result is used in the geodesic equation

$$\frac{d^2x^i}{d\tau^2} + \Gamma^i{}_{kl} \frac{dx^k}{d\tau} \frac{dx^l}{d\tau} = 0 \qquad \left(g_{ik} \frac{dx^i}{d\tau} \frac{dx^k}{d\tau} = 1 \right) \qquad (2.2.2)$$

for the motion in a gravitational field of a test particle with rest mass, or in

$$\delta \int d\tau = 0 \qquad \left(g_{ik} \frac{dx^i}{d\tau} \frac{dx^k}{d\tau} = 0 \right) \qquad (2.2.3)$$

for the motion of rest-massless particles, then one obtains in first approximation, i.e., retaining only terms of the order fM/rc^2, the Kepler motion of the test particles (planets), in the first case, and the gravito-optical effects, in the second case.

The perihelion precession (2.1.8), by contrast, derives from equation (2.2.2) only as a second-order effect and is very sensitive to small (post-Newtonian) changes (2.1.10) in the law of gravitation.

In particular, as was shown in Section 2.1, special-relativistic corrections of this order can be allowed for by small changes of this kind in the law of gravitation. Accordingly, on the one hand, every desired perihelion angle (2.1.8) or (2.1.17) can easily be incorporated in a model of gravitation without us having to dramatically alter the gravitational law; on the other hand, it is difficult to separate the contribution to the perihelion rotation originating in the form of the gravitational law from the contribution due to the kinematic relativistic corrections. The perihelion rotation is therefore, in contrast with the gravito-optical effects of the first order, not a particularly sensitive means for empirically examining a theory of gravitation.

Analyses performed on observations and measurements of gravito-optical effects carried out in recent years have demonstrated that these effects are capable of providing important additional information for experimental gravitational research. Now, in terrestrial experiments, the Einstein slowing down of clocks in a gravitational field has been verified also quantitatively with satisfactory precision. Against that, the amount of the Einstein red shift of the solar spectrum is debatable; an anomalous excess over Einstein's theoretical value might exist. That such an excess is in fact associated with the deflection of light in the solar gravitational field has since Freundlich (1929 and later) repeatedly been claimed; cf. Woodward and Yourgrau (1972).

Remark

It is of special interest to discuss the above-mentioned apparently irreducible anomaly in the observations of light deflection by the sun. Mikhailov (1959), for example, concludes that observations at the solar limb yield, instead of the general-relativistic value of 1.75″, the simple mean value of 2.09″ ± 0.15″ (standard deviation) or the weighted mean value of 1.93″ ± 0.07″; which implies a significant 10% excess over the general relativity theory prediction.

Merat, Pecker, Vigier, and Yourgrau (1974) carried this analysis of the light deflection measurements one step further. They took into consideration that, owing to more or less favorable physical conditions at the time of the solar eclipse and also owing to the differences in techniques and the instruments used in the recording and measurement proceedings, different results were obtained by different observers for the deflection of light by the sun. Merat *et al.* accordingly followed a procedure of attributing weights to each set of observational data, and in this manner found an average excess deflection over the general-relativistic prediction as a function of the distance to the solar limb.

To analytically describe these deflection excesses and the excess red shift (also above the general relativity theory value) in the solar spectrum observed in a variety of experiments [cf. Merat *et al.* (1974)], Treder (1971) introduced an "effective" metric for the spherically symmetric case having the approximate form

$$ds^2 = -\left\{1 + \frac{2fM}{c^2 r} + F_1[u_\nu(r, \nu)]\right\} dl^2$$

$$+\left\{1 - \frac{2fM}{c^2 r} - F_0[u_\nu(r, \nu)]\right\} c^2\, dt^2 \qquad (2.2.4)$$

where $u_\nu = u_\nu(r, \nu)$ denotes the spectral density of the electromagnetic field, while F_0 and F_1 are certain functionals of u_ν. This leads to the effective index of refraction

$$n = \left(-\frac{g_{11}}{g_{00}}\right)^{1/2}$$

$$= \frac{1 + fM/c^2r + \frac{1}{2}F_1}{1 - fM/c^2r - \frac{1}{2}F_0}$$

$$\approx 1 + \frac{2fM}{c^2r} + \frac{1}{2}(F_1 + F_0) \qquad (2.2.5)$$

and the Fermat principle

$$\delta \int n \, dl = 0$$

then predicts, by equation (2.1.69), a light deflection

$$\delta\chi \approx \int_{x^2=-\omega}^{+\omega} \frac{\partial}{\partial x^1}\left(\frac{c}{n}\right) dx^2 \qquad (2.2.6)$$

The red shift now depends on g_{00}; according to equation (2.1.68), it amounts to

$$\frac{\Delta\nu}{\nu} \approx -\frac{fM}{c^2r_0} - \frac{1}{2}F_0[r_0, \nu] \qquad (2.2.7)$$

at the surface $r = r_0$ of the sun.

The predictions of equations (2.2.6) and (2.2.7) agree surprisingly well with the excess deflection and excess red shift if one assumes for F_0 and F_1 a semiempirical function discussed by Merat et al. (1974).

106

Intermezzo: The Einstein Effects

To explain the excesses over the Einstein values of the light deflection and red shift in the solar gravitational field, which he himself claimed to exist, Freundlich (1929) proposed a nonlinear interaction of photons with the electromagnetic radiation field. This hypothesis was supported for a time also by Born in connection with his nonlinear electrodynamics. However, in the framework of special relativity theory alone, the hypotheses of Freundlich and Born on the anomalous Einstein effects appear to be irreconcileable with quantum electrodynamics.

Freundlich's ideas can, however, be brought into harmony with equation (2.2.5) if one assumes that the radiation field operates like a dispersive dielectric, so that the effective speed $c/n(v)$ of a photon with frequency v is a functional of the spectral energy density $u_v(v, x^l)$ of the radiation field. It should still be remarked that one could justify an effective metric as in (2.2.4) with an appeal to the unitary field theories of gravitation and electromagnetism. For, unitary field theories of the same kind as the "asymmetric field theory" of Einstein (1950, 1955) and Schrödinger (1950) (or also the "Hermitian field theory" of Einstein and Straus) contain a dependence of the space–time metric a_{ik} (such that $ds^2 = a_{ik}\,dx^i\,dx^k$) on the energy density u of the electromagnetic radiation field. The concrete form of the functional dependence $a_{ik} = a_{ik}[u, x^l]$ depends, however, partially on the physical interpretation of the antisymmetric part

$$\underset{\vee}{g}_{ik} = \tfrac{1}{2}(g_{ik} - g_{ki})$$

of the unsymmetric fundamental tensor $g_{ik} \neq g_{ki}$ and, above all, it also depends on which symmetric tensor concomitant $a_{ik}[g_{mn}]$ of the g_{mn} is introduced as the metric tensor a_{ik} of space-time (Einstein, 1950, 1955; Hlavatý, 1957; Tonnelat, 1965). The correspondence

107

among different functionals $a_{ik}[g_{mn}]$ and various "effective metrics," one of which corresponds to the pure gravitational metric of general relativity theory, could then also provide clues for the physical interpretation of the Einstein–Schrödinger theories.

Chapter 3

Global Principles and the Theory of Gravitation

3.1. Principles of a Global (Inertia-Free) Theory of Gravitation

In order to attempt a global description of gravitation, one has to consider the entire cosmos, rather than infinitesimal space–time regions. Unfortunately, our experience from nongravitational physics is in the first instance of no use in this approach, as the physics in question has a local, field-theoretical, character. We are left with only one possibility: Start with a global formulation of gravitational law, and afterwards attempt to relate it to local, gravitational as well as nongravitational, physics.

Max Planck (1913) pointed out that the global and the local points of view are not equivalent: The global aspect is always the more general one, in the same sense that "a finite quantity includes an infinitesimal one as a special case." In a similar manner, according to Planck, an instantaneous action-at-a-distance theory is more general than a (local) field theory (an "infinitesimal theory" Planck calls it). Indeed, according to the field-theoretical locality principle, the forces acting at a point depend only on the infinitesimal vicinity of this point, while any instantaneous action-at-a-distance theory holds that these forces are determined by all the other bodies in the universe. Accordingly, Planck concluded that the locality principle, postulating the reducibility of all "telescopic" interactions to "microscopic" ones (Neumann, 1896), implies a simplification of, and also a restriction upon, the nature and mode of action of all natural forces. "Es versteht sich, dass dieser Satz tief eingreift in das Wesen und die Wirkungsweise aller Naturkräfte," Planck declared, in reference to the law asserting the validity of the locality principle.

For Planck, the answer to the question, whether instantaneous action at a distance or field theory is valid, depended on the measure of success achieved by the infinitesimal theory in describing interactions. Considering this question in the context of electrodynamics, Planck (1887) wrote that "at present there is a large degree of probability" for a successful description of all electric phenomena within the field-theoretical framework.

Starting from Planck's observations, we want to add the following two remarks: (1) Despite continued successes of field theories, theoreticians have not succeeded in eliminating all telescopic elements from local, i.e., field, theories. Prominent among these elements are the initial and boundary conditions, which are necessary for the integration of the field equations. (2) The power of the local-theory conception derives possibly from the nature of the interaction. And, since gravitation is the most "telescopic" interaction we know, a breakdown of the local description of interactions is most likely to occur in the area of gravitational interactions.

Global Principles and the Theory of Gravitation

Einstein (1922) maintained—in his polemic with Selety (1922) concerning the relation existing between his own and de Sitter's relativistic world models (de Sitter, 1917) and the Lambert–Charlier "molecular-hierarchic cosmology"—that the "inertia of a single body should be induced, in the same sense as its gravitational force, by the totality of all the other bodies." In this manner Einstein combined his own principle of equivalence between inertia and gravitation with Mach's principle of the relativity of inertia. The resulting principle we call the *Mach–Einstein doctrine*. Einstein rejected the molecular-hierarchic cosmos for the reason that it did not conform with this Mach–Einstein doctrine. Hermann Weyl (1923), an opponent of Mach's principle, on the other hand, regarded the Lambert–Charlier theory of the cosmos as being equivalent to the relativistic cosmology of Einstein and de Sitter.

If we now seek to formulate a global theory of gravitation, it seems clear that we must take as point of departure the principle of equivalence between inertial and heavy masses, the generalized Galilean principle of relativity (according to Poincaré), and the preliminary formulation of the correspondence principle ("Newton's gravitodynamics is valid in any approximation").

First, let us return to the correspondencelike relation between a general theory of gravitation and Newtonian gravitodynamics. This relation requires that small local gravitational effects (with the local potentials $\delta \mid \phi \mid \approx fm/r \ll c^2$) be approximately describable by Newton's equations. As in any local theory of gravitation, particularly general relativity theory, we require moreover that, for a space–time without gravitation, special relativity theory be valid. The transition from the general theory of gravitation to Newton's theory is carried out in a space–time with no gravitational fields other than the weak local ones under consideration. Now, cosmology asserts that c^2 is of the order of the average gravitational

111

potential of a quasi-Einstein cosmos,

$$| \phi | = \frac{fM}{R} \sim c^2$$

(Einstein, 1917). Therefore, one may also require that Newton's theory of gravitation follows for

$$\frac{\delta\phi}{\phi} \approx \frac{m}{r} \frac{R}{M} \to 0 \qquad (3.1.1)$$

Then Newton's theory does not emerge in the transition to field-free space, but rather when $| \phi |$, the average gravitational potential of the universe, becomes very large (Treder, 1972). (Einstein himself regarded the "Einstein cosmos" as an expression of the validity of the Mach–Einstein doctrine in the version formulated more precisely below.)

In view of the foregoing, Newtonian physics should be formulated with reference to the cosmological masses. They, in fact, define a Galilei–Newtonian reference system. This means that the motion of any given particle P_a and especially its acceleration must be described in relation to the total distribution of cosmic particles P_B; moreover, the inertial mass of the particle P_a, which determines the nature of its Galilei–Newtonian inertial motions, must be a function of the Newtonian gravitational potential of all the cosmic mass elements taken together (Mach, 1883; Friedländer, 1896).

This point of view corresponds to a global interpretation of the principle of relativity.

The local versions, discussed in Chapter 1, of a general principle of relativity refer all kinematic and dynamic relations of a particle P_a to the metric field $g_{ik}(x)$ of its infinitesimal neighborhood, an effect one achieves by writing the relations in a generally covariant (Einstein covariant) form. In this case, the general principle of relativity asserts that the dynamic relations of any particle are transformed cogrediently with respect to the metric field in its neighborhood. Therefore, according to the general relativity principle, "relative" here means "referred to the local gravitational field" (Einstein, 1969; Weyl, 1923, 1924).

By contrast, the Galileian position, velocity, and acceleration differences of the particles,

$$\mathbf{r}_{AB} = \mathbf{r}_A - \mathbf{r}_B$$

$$\mathbf{v}_{AB} = \dot{\mathbf{r}}_{AB} = \dot{\mathbf{r}}_A - \dot{\mathbf{r}}_B \qquad (3.1.2)$$

$$\dot{\mathbf{v}}_{AB} = \ddot{\mathbf{r}}_{AB} = \ddot{\mathbf{r}}_A - \ddot{\mathbf{r}}_B$$

are relative quantities with respect to the cosmic masses. Consequently, in this case, the general principle of relativity corresponds to the Poincaré postulate (Poincaré, 1912, 1914): Dynamics does not contain any absolute quantities, but only the differences of the position vectors $\mathbf{r}_A$ and $\mathbf{r}_B$ of the particles P_A and P_B and their time derivatives

$$\frac{d^n \mathbf{r}_{AB}}{dt^n} = \frac{d^n \mathbf{r}_A}{dt^n} - \frac{d^n \mathbf{r}_B}{dt^n} \qquad (3.1.3)$$

Here, relativity has the meaning not of local relativity, but of relativity per distance. This version of relativity can be introduced consistently only if (cf. Neumann, 1870) the kinetic term

$$P = \tfrac{1}{2} \sum_A m_A^* v_A^2 \qquad (3.1.4)$$

113

in the Lagrangian of Newtonian mechanics, which contains the absolute velocities, is replaced by a term involving only the relative velocities $\mathbf{v}_{AB}$. This replacement is possible only if the inertial mass

$$m_a^* = (m_a)_{\text{inertial}}$$

of a particle P_a is reduced to the Newtonian potential function of all the other cosmic particles P_B, such that it becomes a homogeneous linear and scalar (i.e., isotropic) function of the mass potentials fm_B/r_{AB}:

$$m_a^* = -2m_a \frac{\beta}{c^2} \phi$$

$$= \frac{2\beta m_a}{c^2} f \sum_{B \neq a} \frac{m_B}{r_{aB}} \qquad (3.1.5)$$

[cf. Treder (1972, 1974a,b)]. Thus, we have

$$T = \frac{1}{2} \sum_{A>B}^{N} \frac{2\beta}{c^2} f \frac{m_A m_B}{r_{AB}} v_{AB}^2 \qquad (3.1.6)$$

for the kinetic term assigned to the totality of the cosmic particles; in this expression, m_a, m_A, and m_B are heavy masses and β is a numerical constant which is, for cosmological reasons (Einstein cosmos), of the order unity, assuming c to be the speed of light. In this manner, the dynamics of a given particle is formulated with reference to the totality of all the other particles and thus to the rest of the universe (Treder, 1972). Here the isotropy of inertia is given independently of the structure of the universe.

[We should add the remark that instead of T, equation (3.1.6), one could discuss the quantity

$$\bar{T} = \frac{1}{2} \sum_{A>B}^{N} \sum \frac{2\alpha}{c^2} f \frac{m_A m_B}{r_{AB}} \dot{r}_{AB}^2 \qquad (3.1.7)$$

In general, employment of such a function results in an *an*isotropic contribution to inertia. Unlike the Lagrangian formed with T, the Lagrangian arising from the use of $\bar{T}$ satisfies not only the principle of relativity of acceleration but also a principle of relativity for rigid rotations. However, in principle, it makes little sense to look for such a dynamics, because rotational accelerations lead to measurable quantities, i.e., to quantities that are dynamically detectable without the aid of any kinematic reference system. The relativity of rotation required by Mach's principle follows from the relativity of inertia postulated by the Mach–Einstein doctrine (Treder, 1972). Starting from the expression (3.1.7) for $\bar{T}$, one is unable to find any connection with general relativity theory which leads, both for the Einstein cosmos and for the first post-Newtonian approximation of celestial mechanics, to the induction of a scalar inertia (Einstein 1912, 1922). The most general expression for the kinetic energy is

$$T = \frac{1}{2} \sum_{A>B}^{N} \sum \frac{2f}{c^2} \frac{m_A m_B}{r_{AB}} (\alpha \dot{r}_{AB}^2 + \gamma r^2 \dot{\varphi}^2) \qquad (3.1.8)$$

For $\alpha = \beta = \gamma$ this function provides a scalar inertia, and the link with general relativity theory is established if one sets $\beta = \frac{3}{2}$; $\gamma = 0$ implies the identical vanishing of the total angular momentum ("relativity of the rigid rotation"); finally, $\alpha = 0$ gives rise to the vanishing of the total virial and, therefore, to the "relativity of the isotropic dilatation" [(Helmholtz' principle); cf. Treder, 1973a,b.]

Returning now to the representation (3.1.5) of the inertial mass of a particle by its gravitational interaction potential with other cosmic particles, we observe that it contains a consistency condition: Apart from local perturbations, one has, according to Mach's dynamic definition of inertial masses (Mach 1883),

$$m_a^* = m_a \leftrightarrow \frac{2\beta}{c^2} \mid \phi \mid = 1 \qquad (3.1.9)$$

cf. Einstein (1913), Thirring (1921), Dicke (1964), and others. (Comparison with the Einstein cosmos again furnishes $\beta \approx 1$.) The consistency condition (3.1.9) is fulfilled because of the equivalence between the heavy masses m_A and the inertial masses m_I. Therefore, if we understand by the Mach–Einstein doctrine the representation, as in equation (3.1.5), of the inertial mass in the form of a homogeneous linear and scalar function, then this doctrine can be viewed as a mathematical expression of Einstein's equivalence principle. The Mach–Einstein doctrine reduces inertia to Newtonian gravitation. Contrariwise, general relativity theory reduces gravitation to inertia, i.e., to free motion in a metric field. In general relativity theory, both gravitation and inertia belong to the geometric part of Poincaré's epistemological sum,

$$\text{geometry} + \text{interaction} = \text{physics}$$

According to the Mach–Einstein doctrine, on the other hand, both phenomena belong to the interaction term. Classical mechanics occupies an intermediate position: Inertia falls under "geometry" and gravitation under "interaction."

Global Principles and the Theory of Gravitation

The interpretation of the equivalence principle in general relativity theory cannot be distinguished locally from the interpretation given by the Mach–Einstein doctrine; the local inertia and relativistic effects, following, respectively, from general relativity theory and the Mach–Einstein doctrine, must be equivalent. The allusion here is to those effects considered by Einstein (1912; 1913; 1917; 1921), Thirring (1918, 1921), and Fokker (1965), where in general relativity theory the reference system must be chosen such that only real masses contribute to the deformation of the Galilean inertial systems. (For the linearized Einstein equations, this can be done again by choosing the Hilbert gauge which leads to the isotropy of space: $g_{\mu\nu} = \phi\delta_{\mu\nu}$.) The Einstein inertial effects follow from the definition

$$p_\nu = g_{(\nu\nu)}mu^\nu$$

$$\approx \frac{g_{(\nu\nu)}}{(1 - v^2/c^2)^{1/2}(g_{00})^{1/2}} \, mv^\nu \qquad (3.1.10)$$

of momentum (Einstein, 1917); this local induction of inertia corresponds to the Machian induction with $\beta = \frac{3}{2}$.

The number β in equation (3.1.5) is characteristic of the Mach–Einstein universe. We deduced the value $\beta = \frac{3}{2}$ by invoking the principle of the local equivalence of relativistic gravitodynamics and Machian gravitodynamics, a principle originally postulated by Einstein himself. However, the same value of $\beta = \frac{3}{2}$ follows by cosmological arguments from the dependence of the local dynamical quanties on the global properties of the Machian universe: The Machian universe has to correspond to an Einsteinian cosmos as the realization of Mach's principle in general relativity theory; cf. Treder (1974b). The average cosmic gravitational potential ϕ can be written, in terms of an average cosmic distance R and the average mass density ϱ, as

$$\phi = -f \frac{M}{R}$$

$$= -\frac{4\pi}{3} f \varrho R^2 \qquad (3.1.11)$$

But, in the Einstein cosmos of general relativity theory, the equation

$$4\pi f \varrho R^2 = c^2 \qquad (3.1.12)$$

is valid, so that $\phi = -c^2/3$. On combining this result with the Mach principle (3.1.9), the Mach–Einstein numbers

$$\beta = \frac{3}{2}$$

$$\frac{|\phi|}{c^2} = \frac{1}{3} \qquad (3.1.13)$$

are obtained for the Einstein universe.

A theory of gravitation that starts from the Mach–Einstein doctrine leads to a generalized Galilean principle of relativity, which implies a telecomparison and action at a distance. Such a theory of gravitation can therefore not be regarded as a field theory in the sense of special relativity theory. According to the Mach–Einstein doctrine (and also in local relativistic cosmologies), the metric structure of space (and, therefore, its inertial structure) is defined only by the gravitational potential of the cosmic masses. Consequently, the notion of "speed of propagation" is meaningless for the global gravitational field; such a speed is defined only with respect to a given geochronometry. Only for local "perturbations" of the global metric, i.e., for effects embedded in a background that is already defined in the sense of the Mach–Einstein doctrine, will a speed of propagation be definable (and thus a correspondence-like connection with special relativity theory become possible). But if the notion of speed of propagation becomes meaningless, then, according to Einstein, the concept of action at a distance should take its place.

[We note, parenthetically, that in relativistic cosmology, if one assumes the validity of the cosmological principle, the cosmic gravitational potential is also furnished by the Newtonian potential of the fundamental particles that constitute Milne's so-called cosmic cloud. The Friedmann equation of relativistic cosmology asserts the time independence of the Hamiltonian that belongs to this Newtonian N-particle system (McCrea and Milne, 1934; Milne, 1933, 1934, 1948; Heckmann, 1969; Treder, 1973a).]

The geometrization of inertia-free gravitodynamics leads to a geochronometry of space–time which is completely determined by the gravitational potential (as primarily required by Einstein in his *Kosmologische Betrachtungen*). Here, the gravitational action of the cosmic masses is not limited (as in general relativity theory) to modifying the existing ether, that is, the background metric is not independent of the masses and fixed by the boundary conditions (locally as Minkowski or de Sitter space); instead the background metric itself is generated by the masses. According to Einstein's postulate, the spatial metric $g_{\mu\nu}$ becomes a homogeneous linear functional of the gravitational potential of the cosmic masses, which in cosmology is necessarily Newtonian. One has

$$g_{\mu\nu} = g_{\mu\nu}[\phi] \qquad (3.1.14)$$

with

$$g_{\mu\nu} \to 0 \qquad \text{for } \phi \to 0$$

119

For vanishing cosmic masses (no matter anywhere), not only is a flat metric nonexistent, but, in accordance with Einstein's postulate, there is no metric structure at all (or, the metric structure is completely degenerated). Furthermore,

$$\frac{\delta\phi}{\phi} \to 0 \qquad \text{for } \phi \to \infty \tag{3.1.15}$$

3.2. Empirical Consequences of the Relativity of Inertia Principle

In order to discuss some empirical consequences of the relativity of inertia, we consider the Lagrangian (Riemann, 1880; Treder, 1972)

$$L = \sum_{A>B}^{N} \frac{m_A m_B}{r_{AB}} f\left(1 + \frac{\beta}{c^2} v_{AB}^2\right) \tag{3.2.1}$$

and the corresponding Hamiltonian

$$H = -\sum_{A>B}^{N} f \frac{m_A m_B}{r_{AB}} \left(1 - \frac{\beta}{c^2} v_{AB}^2\right) \tag{3.2.2}$$

which follows from the kinetic energy expression (3.1.6).

[Some authors assert that Mach's principle is a selection rule for the boundary conditions in general relativity theory and that this selection rule is satisfied especially by closed relativistic world models with positive space curvature. The Hamiltonian H, equation (3.2.2), shows that the Mach–Einstein doctrine is not connected with the sign of the total energy of the cosmos. For an expanding isotropic Mach–Einstein universe we even have $H \geq 0$, which corresponds relativistically to a flat and a hyperbolic world model (Treder, 1974a).]

We need to assume further on that the universe consists of $N \gg 1$ pointlike particles P_A and that a subsystem consisting of n particles is located approximately in the central region of the "Machian cloud." These assumptions then lead to the result that the expression (cf. Treder, 1972) (Σ' signifies a sum from which the term $A = a$ is deleted):

$$
\begin{aligned}
m_a^* &= \frac{2\beta}{c^2} f m_a \sum_A^N{}' \frac{m_A}{r_{aA}} \\
&= m_a \left[\frac{2\beta}{c^2} |\phi| + \frac{2\beta}{c^2} f \sum_n^n{}' \frac{m_\alpha}{r_{a\alpha}} \right] \\
&\equiv \bar{m}_a^* + \Delta m_a^*
\end{aligned}
\tag{3.2.3}
$$

valid for $n \ll N$, plays the role of the inertial mass of the particle P_a belonging to the n-particle system; herein m_a, m_A, and m_α are the gravitational masses of the particles, while M denotes the mass and R the radius of the cosmic particle cloud. The predominant part $\bar{m}_a^*$ of the mass m_a^* is induced by ϕ, the average of the collective gravitational potential, which may be taken as space-independent at the center of the cloud. Using the appropriate normalization of f, one gets, from equation (3.2.3),

$$
\bar{m}_a^* = m_a \frac{2\beta}{c^2} |\phi| = m_a
\tag{3.2.4}
$$

and

$$
m_a^* = m_a \left(1 + \frac{2\beta}{c^2} f \sum_n^n{}' \frac{m_\alpha}{r_{a\alpha}} \right)
\tag{3.2.5}
$$

In association with the two parts $\bar{m}_a^*$ and $\varDelta m_a^*$ of equation (3.2.3), we may discern two sets of consequences of the relativity of inertia, as described by equation (3.2.3), which are in principle capable of being proved empirically.

The first set comprises consequences that are connected with the space dependence of the expression

$$\varDelta m_a^* = 2\beta m_a \frac{f}{c^2} \sum_{\alpha=1}^{n} \frac{m_\alpha}{r_{a\alpha}} \tag{3.2.6}$$

i.e., with the $r_{\alpha a}$-dependent action of the n-particle system on the particle P_a. For $\beta = \frac{3}{2}$, the expression (3.2.4) corresponds to the dependence of the inertial mass on the local gravitational fields following from general relativity theory.

Accordingly, one finds, for instance, that the inertia of the particle P_a increases when ponderable masses accumulate in its neighborhood. Consider the case of a system consisting of two particles P_1 and P_2, where the particle P_1 is moving around a center occupied by the particle P_2 with mass $\mathfrak{M} \equiv m_2 \gg m_1$. It follows from equation (3.2.5) (with $\beta = \frac{3}{2}$), for $v \ll c$ and $f\mathfrak{M}/r \ll c^2$, that P_1 describes a Kepler ellipse about P_2 that is characterized by an Einsteinian perihelion advance

$$\delta\psi = 3 \frac{2\pi f\mathfrak{M}}{ac^2(1 - e^2)}$$

This result shows the compatibility of general relativity theory and the Mach–Einstein doctrine. Indeed, the Riemann potential in equation (3.2.1) is a special post-Newtonian potential, as discussed above (Treder, 1975).

The second set of consequences, stemming from the relativity of inertia, concerns the term

$$\bar{m}_a^* = \frac{2\beta m_a}{c^2} \mid \phi \mid \tag{3.2.7}$$

Neglecting the local mass induction, one may write, according to (3.1.9),

$$m_a^* = 3m_a \frac{\mid \phi \mid}{c^2}$$

$$= m_a \frac{3fM}{c^2R} \tag{3.2.8}$$

where the cosmological radius R corresponds to the radius of the Einstein cosmos. This formula predicts the existence of a secular effect and thus opens up possibilities for an experimental decision on the validity of the Mach–Einstein doctrine.

Possible experimental tests of the right kind present themselves within our earth–moon system, as inertial effects determine both the shape of the earth and the orbit of our moon (Treder, 1972): (1) The Mach–Einstein doctrine predicts a secular variation of the earth's oblateness of the geophysically required order (a decrease of about 1% per 10^8 years), although one should be aware that an additional effect might result from tidal friction. (2) The Mach–Einstein doctrine further requires the occurrence of a real secular acceleration of the moon. (3) This effect, in turn, gives rise to a secular decrease of the earth–moon distance of about 4 cm per year. Because both the Dirac–Jordan hypothesis (Jordan, 1961) and Darwin's theory of tidal deceleration predict an increase of the earth–moon separation, accurate measurement of this distance over many years could distinguish experimentally between the Dirac–

123

Jordan hypothesis and the Darwin theory, on the one hand, and the Mach–Einstein doctrine, on the other.

Finally, we mention the class of effects that are consequences of the relativity of accelerations, i.e., of the occurrence of relative accelerations in the equations of motion. In general relativity theory, these effects reveal themselves through the form of the Einstein–Thirring terms in the equations of motion (Einstein, 1913; Thirring, 1918, 1921); they correspond to a relativization of d'Alembert's apparent forces of classical dynamics. These effects become significant in connection with large nonisotropic relative motions in the universe.

In carrying out the geometrization of inertia-free mechanics in Hertz's configuration space (Hertz, 1894), it may be noted, one treats the Einstein–Thirring effects as consequences of the projection from the spaces V_{3N} and V_{4N} into the usual Riemannian spaces V_3 and V_4, respectively. Consequently, the appearance of these effects demonstrates the "physical reality" of the relativized Hertzian configuration space. [Geometrized inertia-free dynamics in Hertzian configuration space is studied in detail in a monograph by Treder (1974b).]

If the particle cloud whose dynamics is determined by the Lagrangian (3.2.1) is viewed as a realistic model of the universe, then our universe will necessarily expand and its radius R thus will become a function of the time. Accordingly, one gets for the inertial masses m^* at the time t (Treder, 1972)

$$m^*(t) = \frac{R(t_0)}{R(t)}\, m^*(t_0)$$

$$= 3\,\frac{fM}{c^2 R(t)}\, m \tag{3.2.9}$$

On defining the instantaneous values m^*, according to Mach, by

$$m^* = m \qquad (3.2.10)$$

the effective gravitational constant, determined by Newton's gravitational law, is seen to obey

$$f(t) = f(t_0) \frac{R(t)}{R(t_0)}$$

$$= \frac{1}{3} \frac{c^2}{M} R(t) \qquad (3.2.11)$$

The same kind of proportionality to $R(t)$ is valid for any other physical coupling "constant" $k(t)$, i.e., the ratio $f(t)/k(t)$ is time-independent. As a consequence, cosmological effects result from the Mach–Einstein doctrine only for those processes in which inertia does not play any role.

A Mach–Einstein universe, as here described, is characterized by the fact that "laboratory" measurements of local physical quantities imply statements about macroscopic quantities that relate to the universe as a whole. Thus, in the Mach–Einstein universe, there exist relations among the microscopic quantities of elementary particle physics, on the one hand, and the macroscopic constants that determine the "size of the world," on the other. Such relations have been postulated repeatedly ever since the appearance of Eddington's *Fundamental Theory* (Eddington, 1936, 1948); Eddington, Whittaker, and Heisenberg all viewed them as supplying both a cosmological foundation for quantum and elementary particle physics and a quantum-physical foundation for cosmology. On the other hand, in keeping with the ideas of Milne and Dirac (cf. Jordan, 1961), the dependence of Eddington's cosmic numbers and the universal constants on the age of the universe has also often been reiterated. Both Milne and Eddington, it should be noted, referred to Mach's principle. And the discussion of equations (3.2.7)–(3.29) shows (cf. Treder, 1972) that the conceptions of Eddington and Heisenberg can be reconciled with one another.

The strong influence of the entire universe on local physics and on the physics of finite parts (subsystems) of the universe, which is implied by the assumption of a Mach–Einstein universe, also affects the evolution of large cosmic objects.

3.3. Expansion and Angular Momentum of Large Cosmic Masses

According to both Newtonian and local-relativistic gravitational theories, a massive or a very dense cosmic object will collapse, giving rise to a so-called "black hole" in space. It is further predicted that the structure at the center of the black hole will be unstable and thus continue to contract, until, after a finite but very short interval of proper time, all the matter is concentrated (assuming spherical symmetry of the object) in a single point.

By contrast, stability (or metastability) of large masses with nuclear densities becomes theoretically possible in gravitational dynamic theories which (1) either contain an "absorption of gravity" through a dependence of the effective gravitational constant f^* on the local gravitational potential ϕ (the symbol χ denotes a pure number):

$$f^* = f\left(1 - \chi \frac{|\phi|}{c^2}\right) \qquad (3.3.1)$$

[this is the case in the reference tetrad theory of gravitation; see equations (1.4.11) ff.], or which (2) yield, reciprocally, an increase of the effective inertial mass m^* as a result of an induction of inertia through the local gravitational potential:

$$m^* = \left(1 + 2\beta \frac{|\phi|}{c^2}\right)m \qquad (3.3.2)$$

In what follows we discuss the behavior of large cosmic masses in a Mach–Einstein universe for which the inertial masses are linear functions of the gravitational potential. These large masses will serve at the same time as models of the metagalaxy, i.e., the total universe, which is plainly the largest possible mass.

By a "large cosmic mass" we will, to be precise, understand a particle system of mass $\mathfrak{M}$ whose gravitational radius

$$r_{\text{grav}} = a = \frac{f\mathfrak{M}}{c^2} \qquad (3.3.3)$$

is much larger than its "baryon radius" b. To define this radius, we imagine that the mass $\mathfrak{M}$ is composed of N neutrons of mass μ; thus

$$\mathfrak{M} = N\mu \qquad (3.3.4)$$

The baryon radius

$$b \approx N^{1/3} \frac{h}{\mu c} \qquad (3.3.5)$$

where h is Planck's constant and $h/\mu c$ the Compton wavelength of the neutron, is then approximately the radius which the system would have if it were compressed to nuclear densities. For a large mass $\mathfrak{M}$ one therefore finds the estimate

$$a = \frac{f\mathfrak{M}}{c^2} \gg b \approx \frac{N^{1/3}h}{\mu c} \approx \frac{\mathfrak{M}^{1/3}h}{\mu^{4/3}c} \qquad (3.3.6)$$

which leads to the mass and baryon number estimates

$$\mathfrak{M} \gg \left(\frac{c^3 h^3}{f^3}\right)^{1/2} \frac{1}{\mu^2} \approx 10^{34}\,\mathrm{g} \qquad (3.3.7)$$

and

$$N \gg \left(\frac{c^3 h^3}{f^3}\right)^{1/2} \frac{1}{\mu^3} \approx 10^{58} \qquad (3.3.8)$$

Equations (3.3.6)–(3.3.8) are always fulfilled for galaxies as well as for clusters of galaxies.

The above definition of a large cosmic mass implies that if its material constituents are packed so tightly that they form what may be described as a superneutron star or a "hypernucleus," consisting of nuclear matter (Ambartsumjan, 1975), then the system under consideration will be concentrated far within its gravitational radius.

Consider now a homogeneous particle cloud of total mass $\mathfrak{M}$, which we imagine as being made up of N neutrons or n equal particles of mass m, so that

$$\mathfrak{M} = N\mu = nm \qquad (3.3.9)$$

Let r denote the instantaneous radius of this cloud, that is, a quantity equal in order of magnitude to the average separation between two particles of the cloud. Then it follows, from equation (3.2.1), that the cloud's gravitational dynamics is described by (the dot indicates differentiation with respect to time):

$$L = \frac{f \mathfrak{M}^2}{r} \left(1 + \beta \frac{v^2}{c^2}\right) + \frac{\mathfrak{M}}{2} v^2$$

$$= \frac{f \mathfrak{M}^2}{r} \left[1 + \frac{\beta}{c^2} (\dot{r}^2 + r^2 \dot{\varphi}^2)\right] + \frac{\mathfrak{M}}{2} (\dot{r}^2 + r^2 \dot{\varphi}^2) \qquad (3.3.10)$$

where a rotation of the system has been introduced through the mean angular velocity $\dot{\varphi}$, defined in accordance with

$$v^2 = \dot{\mathbf{r}}^2 = \dot{r}^2 + r^2 \dot{\varphi}^2$$

Inspection of equation (3.3.10) reveals that, unlike the situation in Newtonian dynamics, the internal kinetic energy of the cloud is given by

$$T = \frac{1}{2} \mathfrak{M} \left(1 + 2\beta \frac{f \mathfrak{M}}{c^2 r}\right)(\dot{r}^2 + r^2 \dot{\varphi}^2)$$

$$= \frac{1}{2} \mathfrak{M}^* \dot{\mathbf{r}}^2 \qquad (3.3.11)$$

here

$$\mathfrak{M}^* = \mathfrak{M} \left(1 + 2\beta \frac{f \mathfrak{M}}{r c^2}\right) \qquad (3.3.12)$$

signifies the effective inertial mass of the cloud relative to its internal interactions, which plainly means that the inertial masses of the cloud particles are furnished by

$$m^* = m\left(1 + 2\beta\,\frac{f\,\mathfrak{M}}{c^2 r}\right)$$

(3.3.13)

$$\mathfrak{M}^* = nm^*$$

The last formula leads to the conclusion that the effective mass m^* tends to infinity when the cloud radius r tends to zero. This circumstance actually prevents, as will be shown, a gravitational collapse of the kind encountered in relativistic astrophysics. One finds rather that, for $\beta > 1$, the radial velocity $\dot{r}$ remains finite and less than c as $r \to 0$ (see below).

The energy integral corresponding to the Lagrangian (3.3.10) has the form

$$H = -\frac{f\,\mathfrak{M}^2}{r}\left(1 - \frac{\beta}{c^2}\,(\dot{r}^2 + r^2\dot{\varphi}^2)\right) + \frac{\mathfrak{M}}{2}\,(\dot{r}^2 + r^2\dot{\varphi}^2)$$

$$= \text{const}$$

(3.3.14)

while the integral of the angular momentum reads

$$J = \frac{2\beta}{c^2}\,f\,\mathfrak{M}^2 r\dot{\varphi} + \mathfrak{M} r^2\dot{\varphi} = \text{const}$$

(3.3.15)

Together, these two expressions determine the cloud dynamics. Consider now the cloud in a highly condensed state, such that

$$r \approx b \ll \frac{f\,\mathfrak{M}}{c^2}$$

Then the foregoing integrals reduce to, respectively,

$$H = -\frac{f\mathfrak{M}^2}{r}\left[1 - \frac{\beta}{c^2}(\dot{r}^2 + r^2\dot{\varphi}^2)\right]$$ (3.3.16)

and

$$J = \frac{2\beta}{c^2}f\mathfrak{M}^2 r\dot{\varphi}$$ (3.3.17)

(which implies $r\dot{\varphi} = k = \text{const}$), so that, on elimination of $\dot{\varphi}$, one obtains the Hamiltonian

$$H = -\frac{f\mathfrak{M}^2}{r} + \frac{\beta}{c^2}\frac{f\mathfrak{M}^2\dot{r}^2}{r} + \frac{J^2 c^2}{4\beta f\mathfrak{M}^2 r}$$ (3.3.18)

Differentiation of this function leads, for $r \ll f\mathfrak{M}/c^2$, to

$$\ddot{r} = \frac{Hc^2}{2\beta f\mathfrak{M}^2}$$ (3.3.19a)

and thus to the equation

$$r = \frac{Hc^2 t^2}{4\beta f\mathfrak{M}^2} + At + B$$ (3.3.19b)

(A and B constants) for the radial motion of the cloud. For a positive energy integral ($H > 0$), these results describe a cloud expansion with linearly increasing velocity. The case $H < 0$ is clearly physically impossible, since it would give rise to negative values of r for large enough times t.

The connection between the energy constant H and the angular momentum constant J follows from the assumed initial state of the system: Let, at time $t = 0$, the system be in a metastable state defined by

$$r(0) = r_0 \approx b$$

$$\dot{r}(0) = 0 \qquad\qquad (3.3.20)$$

Substitution of these conditions into equation (3.3.19b) gives

$$A = 0$$

$$B = r_0 \approx \beta \qquad\qquad (3.3.21)$$

while substitution into equation (3.3.17) leads to the desired connection

$$J^2 = 4\beta \frac{f^2 \mathfrak{M}^4}{c^2} + 4\beta \frac{f \mathfrak{M}^2}{c^2} H r_0 \qquad\qquad (3.3.22a)$$

or approximately, for $r_0 \ll f\mathfrak{M}/c^2$,

$$J \approx 2\beta^{1/2} \frac{f \mathfrak{M}^2}{c} \qquad\qquad (3.3.22b)$$

For finite r_0 and $H > 0$, the angular momentum J according to (3.3.22a) is always somewhat larger than the approximate value (3.3.22b). The relative error, for $r_0 \approx b \sim \mathfrak{M}^{1/3}$ and $H \sim \mathfrak{M}$ [see below, equation (3.3.39)], is proportional to $\mathfrak{M}^{-2/3}$, which means that equation (3.3.22b) is a better approximation the larger the system.

The value of the energy constant H follows from the energy law (3.3.14) for large values of r. With the aid of equation (3.3.15), and assuming $r \gg f\mathfrak{M}/c^2$, one finds approximately

$$H \approx \frac{\mathfrak{M}}{2}\,\dot{r}^2 + \frac{J^2}{2\mathfrak{M}r^2} \qquad (3.3.23)$$

and, asymptotically,

$$H = \frac{\mathfrak{M}}{2}\,\dot{r}^2 \qquad \text{for } r \to \infty \qquad (3.3.24)$$

In order to determine the limiting values of $\dot{r}$ for $r \to \infty$, we observe that, with the increase of time, the radial expansion $r(t)$ of the system must go over into the cosmological expansion of the metagalaxy. Accordingly, we identify, for $r \gg f\mathfrak{M}/c^2$, the velocity $\dot{r} \approx v$ with the expansion velocity $\dot{R}$ of the metagalaxy. If $R(t)$ denotes the radius (or distance factor) of the metagalaxy, then it is required that

$$\dot{r} \to \dot{R} \qquad \text{for } t \to \infty \Rightarrow \dot{r} \to \infty \Rightarrow H = \frac{\mathfrak{M}}{2}\,\dot{R}^2 \qquad (3.3.25)$$

where $\dot{R}/R$ is Hubble's constant.

From the dynamics of an isotropic expanding cosmos it now follows generally (Treder, 1976a) that it is possible to introduce additional masses into the cosmos only if the "total energy" $\mathfrak{H}$ of the cosmological expansion, i.e., the energy constant in the equation of motion for $R(t)$ (e.g., in the Friedmann equation of relativistic cosmology) happens to vanish. As is known, $\mathfrak{H} = 0$ leads in both Newtonian and relativistic cosmologies to a flat Einstein–de Sitter cosmos for which

$$R \propto (t - t_0)^{2/3}$$

is an integral of Friedmann's equation. However, in the inertia-free dynamics with Riemannian gravitational potential for a Mach–Einstein cosmos, Friedmann's equation is replaced by a differential equation of the form (3.3.16) (Treder, 1974a)

$$-\frac{fM^2}{R}\left(1 - \beta\frac{\dot{R}^2}{c^2}\right) = \mathfrak{H} \geq 0 \qquad (\beta = \tfrac{3}{2}) \qquad (3.3.26)$$

For $\mathfrak{H} = 0$, this yields the linear expansion law

$$R = \frac{C}{\beta^{1/2}}\, t + \text{const} \qquad (3.3.27)$$

[For a Mach–Einstein cosmos (3.2.10), the energy law asserts

$$0 = \dot{\mathfrak{H}} = -\mathfrak{H}\frac{\dot{R}}{R} + \frac{2f\beta}{c^2} M^2 \frac{\dot{R}}{R} \ddot{R} + 2\mathfrak{H}\frac{\dot{M}}{M}$$

This equation shows that an arbitrary mass $\Delta M = m$ can be added to the cosmos without changing its dynamics when $\mathfrak{H}$ vanishes. Indeed, one then gets identically, for each particle in the cosmos,

$$\frac{mfM}{R} = m\frac{f\beta}{c^2}\frac{M}{R}\dot{R}^2$$

that is, the potential and kinetic energies of every particle add up to zero.]

By equation (3.3.27), the asymptotic fitting of the expanding particle cloud of mass $\mathfrak{M}$ to Riemannian gravitodynamics in the

Mach–Einstein expanding cosmos requires that, for $r \rightarrow \infty$,

$$\dot{r} \rightarrow \dot{R} = \frac{c}{\beta^{1/2}}$$

and thus

$$H = \frac{\mathfrak{M}}{2} \dot{r}^2 = \frac{\mathfrak{M}}{2} \dot{R}^2 = \frac{\mathfrak{M}c^2}{2\beta} \qquad (3.3.28)$$

which determines the energy constant H in equation (3.3.14).

Summarizing our results, we arrive at a mathematical model for an expanding cosmic system, of very large mass $\mathfrak{M}$, that is based on the Lagrangian

$$L = \frac{f\mathfrak{M}^2}{r} \left(1 + \frac{3}{2} \frac{v^2}{c^2} \right) + \frac{\mathfrak{M}}{2} v^2 \qquad (3.3.29)$$

the energy integral

$$H = -\frac{f\mathfrak{M}}{r} \left(1 - \frac{3}{2} \frac{v^2}{c^2} \right) + \frac{\mathfrak{M}}{2} v^2$$

$$= \frac{\mathfrak{M}}{3} c^2 \qquad (3.3.30)$$

and the approximate angular momentum

$$J = \frac{3f\mathfrak{M}^2}{c^2} r\dot{\varphi} + \mathfrak{M}r^2\dot{\varphi}$$

$$\approx 6^{1/2} \frac{f\mathfrak{M}^2}{c} \left(1 + \frac{Hr_0}{2f\mathfrak{M}^2} \right)$$

$$\approx 6^{1/2} \frac{f\mathfrak{M}^2}{c} + \frac{1}{6^{1/2}} N^{4/3}h \qquad (3.3.31)$$

where in the last step we made use of equation (3.3.30) and substituted for r_0 its value according to equation (3.3.31).

Let the expansion of the cosmic system commence at time $t = 0$ when its radius equals its baryon radius b:

$$r(0) = r_0 = b = N^{1/3} \frac{h}{\mu c}$$

$$\approx \mathfrak{M}^{1/3} \frac{h}{\mu^{4/3} c} \ll \frac{f \mathfrak{M}}{c^2} \qquad (3.3.32)$$

Initially the expansion velocity $\dot{r}(t)$ increases linearly, and for $r \ll f\mathfrak{M}/c^2$ one obtains

$$r \approx \frac{Hc^2}{12f\mathfrak{M}} t^2 + b$$

$$\approx \frac{c^4}{36f\mathfrak{M}} t^2 + b \qquad (3.3.33)$$

But in the region $r \approx f\mathfrak{M}/c^2$ the acceleration $\ddot{r}$ of the radial motion begins to diminish, allowing the expansion velocity to asymptotically attain its maximum value in the region $r \gg f\mathfrak{M}/c^2$:

$$\dot{r} = \left(\frac{2}{3}\right)^{1/2} c \qquad \text{for } t \to \infty, \ r \to \infty \qquad (3.3.34)$$

[Equation (3.3.33) would incorrectly predict $\dot{r} \geq c$ for $r \gtrsim 2\beta^2 f\mathfrak{M}/c^2 = 9f\mathfrak{M}/2c^2$.] In the metastable state that precedes the expansion phase, i.e., for $t < 0$, the large mass $\mathfrak{M}$ forms a "hypernucleus" (or, if one prefers, a superheavy elementary particle) with

$$\mathfrak{M} = N\mu \approx N \cdot 1.7 \times 10^{-24} \text{ g} \qquad (3.3.35)$$

and the spin

$$J \approx 6^{1/2} \frac{f \mathfrak{M}^2}{c}$$

$$\approx 6^{1/2} N^2 \frac{f \mu^2}{c}$$

$$\approx N^2 \cdot 10^{-65} \text{ g cm}^2 \text{ sec}^{-1}$$

$$\gg N^{4/3} h \qquad (\text{for } N \gg 1) \tag{3.3.36}$$

corresponding to the substitution $k = r_0 \dot{\varphi} \approx b \dot{\varphi} \approx (2/3)^{1/2} c$ in equation (3.3.31).

It is rather interesting and, in view of Ambartsumjan's (1975) conception of the evolution of compact galaxies (or compact clusters of such galaxies), also very satisfying that the relations (3.3.35) and (3.3.36) reproduce approximately the asymptotic shape of a Regge trajectory for hadrons; cf. Muradjan (1975). Our mathematical model of a radially expanding system, characterized by a large mass and an initially nuclear density $[r(0) \approx b]$, is thus seen to provide a simple dynamical interpretation of Ambartsumjan's (1975) cosmogonic hypothesis concerning the origin of galaxies or clusters of galaxies.

It is evident (Ambartsumjan, 1975) that Ambartsumjan's understanding of the evolution of galaxies is irreconcilable with the principles of Newton's and Einstein's gravitational theories, because, according to both these theories, the internal gravitational potential of a system grows without limit as its mass and density increase, and this leads inevitably to a final state of collapse. By contrast, according to Riemannian gravitational dynamics (and in agreement with the Mach–Einstein doctrine), the ratio of the inertial mass to the gravitational mass grows with increasing local gravitational potential in such a manner that there exists a finite limit for the gravitational acceleration inside very large and very dense systems.

The foregoing is reflected in our model by the following circumstance: In the expression (3.3.18) for the energy H, with $\dot{r} = 0$, the negative of the gravitational energy, $f\mathfrak{M}^2/r_0$, admittedly grows without bound for diminishing radius r_0, but the same is true for the kinetic energy T (as determined by the angular momentum J). Indeed, given a finite rotational velocity $v_0 = r_0\dot{\varphi} = k = c/\beta^{1/2}$, the dependence of T on r_0 is furnished by

$$T = \frac{f\mathfrak{M}^2}{r_0} c^2$$

$$= \frac{J^2 c^2}{4\beta f \mathfrak{M}^2 r_0}$$

for vanishing r_0. The system is stable for $r_0 \to 0$ if J has the finite value

$$J = 2\beta^{1/2} \frac{f\mathfrak{M}^2}{c}$$

$$= 2\beta \frac{f\mathfrak{M}^2}{c^2} r_0\dot{\varphi} \qquad\qquad (3.3.37)$$

and the system must expand, once it has been disturbed, provided

$$J \geq 2\beta^{1/2} \frac{f\mathfrak{M}^2}{c} \qquad \text{for } r_0 > 0 \qquad\qquad (3.3.38)$$

The energy H of the system is positive if equation (3.3.37) is fulfilled. However, since equation (3.3.38) can be viewed as the

condition for the primary galactic nuclei to lie on a Regge trajectory, the latter therefore guarantees $H > 0$. From the asymptotic
behavior of the system it then follows that

$$H = \tfrac{1}{3} \mathfrak{M} c^2$$

$$= \tfrac{1}{3} \mathfrak{M} \mid \phi \mid \qquad (3.3.39)$$

3.4. Inertia-Free Dynamics and Hamilton's Equations of Motion

Finally, we want to present the Hamiltonian (canonical) form
of inertia-free Machian dynamics (Treder, 1976b; cf. also Yourgrau
and Mandelstam, 1960).

Because the Lagrangian appropriate to this dynamics has the
form (3.2.1), a canonical momentum*

$$\mathbf{p}_A = \frac{\partial L}{\partial \mathbf{v}_A}$$

$$= \sum_B{}' \frac{\partial L}{\partial \mathbf{v}_{AB}}$$

$$= \frac{2\beta}{c^2} \, f m_A \sum_B{}' \frac{m_B \mathbf{v}_{AB}}{r_{AB}} \qquad (3.4.1)$$

belongs to each particle A, where $\sum_B{}' = \sum_{B \neq A}$. In combination

---

* Throughout the present section we shall find it convenient to use a
notation according to which $\partial L / \partial \mathbf{v}$, for example, stands for the par-
ticle vector with components $\partial L / \partial \dot{x}$, $\partial L / \partial \dot{y}$, $\partial L / \partial \dot{z}$.

with the Mach–Einstein condition (3.1.9), this means that

$$\mathbf{p}_A = m_A \mathbf{v}_A - \frac{2\beta}{c^2} f m_A \sum_B' \frac{m_B}{r_{AB}} \mathbf{v}_B \qquad (3.4.2)$$

The center of gravity of the cosmos "is at rest in every system of reference":

$$\sum_A \mathbf{p}_A = \sum_A \sum_B' \frac{\partial L}{\partial \mathbf{v}_{AB}} = 0 \qquad (3.4.3)$$

Like the aether in special relativity theory, it therefore has no defined state of motion at all. The equations of motion for the particles A read

$$\dot{\mathbf{p}}_A = \frac{d}{dt} \frac{\partial L}{\partial \mathbf{v}_A} = \frac{\partial L}{\partial \mathbf{r}_A} = \sum_B' \frac{\partial L}{\partial \mathbf{r}_{AB}}$$

$$= -f m_A \sum_B' \frac{m_B}{r_{AB}^3} \mathbf{r}_{AB} \left(1 + \beta \frac{v_{AB}^2}{c^2}\right) \qquad (3.4.4)$$

According to Riemann's extension of Lagrange's theorem (see Treder, 1972), the energy integral belonging to L is

$$E = \sum_A \mathbf{v}_A \cdot \frac{\partial L}{\partial \mathbf{v}_A} - L$$

$$= \sum_{A>B} \sum \mathbf{v}_{AB} \cdot \frac{\partial L}{\partial \mathbf{v}_{AB}} - L$$

$$= -\sum_{A>B}^{N} \sum f \frac{m_A m_B}{r_{AB}} \left(1 - \frac{\beta}{c^2} v^2\right) = \text{const} \qquad (3.4.5)$$

It equals the Hamiltonian (3.2.2) of the system. (In what follows we put light speed $c = 1$.)

The essential features of the Hamiltonian form of inertia-free dynamics already make their appearance in the two-particle problem ($N = 2$), which we now consider.

By equation (3.2.1), the Lagrangian appropriate to a system consisting of the particles 1 and 2 is

$$\mathscr{L}(\mathbf{r}, \mathbf{v}) = \frac{f\mu M}{r}\,(1 + \beta v^2) \qquad (3.4.6)$$

wherein M and μ denote the gravitational masses of the two particles, $\mathbf{r} = \mathbf{r}_1 - \mathbf{r}_2$ their vector separation, and $\mathbf{v} = \mathbf{v}_1 - \mathbf{v}_2$ their relative velocity. The corresponding canonical momenta are

$$\mathbf{p}_1 = -\mathbf{p}_2 = \frac{\partial \mathscr{L}}{\partial \mathbf{v}} = 2\beta\,\frac{f\mu M}{r}\,\mathbf{v} \qquad (3.4.7)$$

The equations of motion determine solely the relative motion of particles 1 and 2; with $\mathbf{p} \equiv \mathbf{p}_1$, we get

$$\dot{\mathbf{p}} = \frac{d}{dt}\,\frac{\partial \mathscr{L}}{\partial \mathbf{v}} = \frac{\partial \mathscr{L}}{\partial \mathbf{r}} \qquad (3.4.8a)$$

and, therefore,

$$2\beta f\,\frac{\mu M}{r}\left(\dot{\mathbf{v}} - \mathbf{v}\,\frac{\dot{r}}{r}\right) = -\,\frac{f\mu M}{r^3}\,\mathbf{r}(1 + \beta v^2) \qquad (3.4.8b)$$

The energy integral becomes

$$E = \mathbf{v} \cdot \frac{\partial \mathscr{L}}{\partial \mathbf{v}} - \mathscr{L}$$

$$= -\frac{f\mu M}{r}(1 - \beta \mathbf{v}^2)$$

$$= \text{const} \tag{3.4.9}$$

which, on elimination of $\mathbf{v}$ in favor of $\mathbf{p} = \mathbf{p}_1$, as defined in (3.4.7), leads to the Hamiltonian

$$\mathfrak{H}(\mathbf{r}, \mathbf{p}) = E = -\frac{f\mu M}{r} + \frac{rp^2}{4f\mu M\beta} \tag{3.4.10}$$

[Interesting here is the completely different dependence on r of the functions $\mathfrak{H}(\mathbf{r}, \mathbf{p})$ and $\mathscr{L}(\mathbf{r}, \mathbf{v})$.] Differentiation of this function gives rise to Hamilton's canonical equations of motion:

$$\mathbf{v} = \dot{\mathbf{r}} = \frac{\partial \mathfrak{H}}{\partial \mathbf{p}}$$

$$= \frac{2r}{4\mu M f\beta} \mathbf{p} \tag{3.4.11}$$

and

$$-\dot{\mathbf{p}} = \frac{\partial \mathfrak{H}}{\partial \mathbf{r}}$$

$$= \frac{f\mu M}{r^3}\mathbf{r} + \frac{p^2}{4\mu M f\beta}\frac{\mathbf{r}}{r}$$

$$= \frac{f\mu M}{r^3}\mathbf{r}(1 + \beta v^2) \tag{3.4.12}$$

the last result, of course, simply recovering (3.4.8), by virtue of
$\partial \mathscr{L}/\partial \mathbf{r} = \dot{\mathbf{p}} = -\partial \mathfrak{H}/\partial \mathbf{r}$.

The Lagrangian (3.4.6) furnishes the virial

$$V = \mathbf{r} \cdot \mathbf{p} = \mathbf{r} \cdot \frac{\partial \mathscr{L}}{\partial \mathbf{v}} = 2\beta f\mu M \dot{r} \qquad (3.4.13)$$

with $\dot{r} = \mathbf{v} \cdot (\mathbf{r}/r)$, for which, with the aid of the definition (3.4.6) and the equation of motion (3.4.12), one can establish the virial theorem

$$\dot{V} = \dot{\mathbf{r}} \cdot \mathbf{p} + \mathbf{r} \cdot \dot{\mathbf{p}}$$

$$= \dot{\mathbf{r}} \cdot \frac{\partial \mathscr{L}}{\partial \mathbf{v}} + \mathbf{r} \cdot \frac{d}{dt}\left(\frac{\partial \mathscr{L}}{\partial \mathbf{v}}\right)$$

$$= \frac{2f\mu M}{r} \beta v^2 - \frac{f\mu M}{r^3} \mathbf{r}^2(1 + \beta v^2)$$

$$= -\frac{f\mu M}{r}(1 - \beta v^2)$$

$$= E = \text{const} \qquad (3.4.14)$$

The post-Newtonian two-body problem with $N \gg 1$ includes, as a limiting case, the motion of a test particle $m_1 = \mu$ in the central field of a mass $\mathfrak{M}$ and, therefore, the equivalent of the relativistic Kepler problem (Treder, 1972).

The appropriate Lagrangian,

$$l = \mu \frac{v^2}{2} + \frac{f\mu \mathfrak{M}}{r}\left(1 + \beta \frac{v^2}{c^2}\right) \qquad (3.4.15)$$

yields the canonical momentum of the test particle:

$$\mathbf{p} = \frac{\partial l}{\partial \mathbf{v}}$$

$$= 2\mu\left(\frac{1}{2} + \frac{f\beta}{c^2}\,\frac{\mathfrak{M}}{r}\right)\mathbf{v} \tag{3.4.16}$$

Accordingly, the Hamiltonian becomes

$$h(\mathbf{r}, \mathbf{p}) = \mathbf{v} \cdot \frac{\partial l}{\partial \mathbf{v}} - l$$

$$= \frac{\mu}{2}\,v^2 - \frac{f\mu\mathfrak{M}}{r}\left(1 - \beta\,\frac{v^2}{c^2}\right)$$

$$= \mu\left(\frac{1}{2} + \frac{\beta f}{c^2}\,\frac{\mathfrak{M}}{r}\right)v^2 - \frac{f\mu\mathfrak{M}}{r}$$

$$= \frac{p^2}{4\mu\left(\dfrac{1}{2} + \dfrac{\beta f}{c^2}\,\dfrac{\mathfrak{M}}{r}\right)} - \frac{f\mu\mathfrak{M}}{r} \tag{3.4.17}$$

The Lagrange equation of motion reads

$$\dot{\mathbf{p}} = \frac{d}{dt}\,\frac{\partial l}{\partial \mathbf{v}} = \frac{\partial l}{\partial \mathbf{r}}$$

$$= -\frac{\mathbf{r}}{r^3}\,f\mu\mathfrak{M}\left(1 + \beta\,\frac{v^2}{c^2}\right) \tag{3.4.18}$$

which is the same as Hamilton's equation of motion

$$-\dot{\mathbf{p}} = \frac{\partial h}{\partial \mathbf{r}} = \frac{\mathbf{r}}{r^3}\,f\mu\mathfrak{M}\left[1 + \frac{p^2}{4\mu}\left(\frac{1}{2} + \frac{f\mathfrak{M}}{r}\,\frac{\beta}{c^2}\right)^{-2}f\mathfrak{M}\beta\right]$$

$$= \frac{\mathbf{r}}{r^3}\,f\mu\mathfrak{M}\left(1 + \beta\,\frac{v^2}{c^2}\right) \tag{3.4.19}$$

The remaining Hamilton's equation of motion has the form

$$\mathbf{v} = \frac{\partial h}{\partial \mathbf{p}}$$

$$= \mathbf{p}\left(\mu + \frac{f\mu\mathfrak{M}}{r}\ \frac{2\beta}{c^2}\right)^{-1} \tag{3.4.20}$$

To complete our analysis, we present the canonical equations of motion for the N-particle cosmos. Towards this end, it is convenient to define the relative quantities (Treder, 1974b)

$$m_{AB} = \frac{2\beta}{c^2}\ \phi_{AB}$$

$$= \frac{2\beta f}{c^2}\ \frac{m_A m_B}{r_{AB}} \tag{3.4.21}$$

and

$$p_{AB} = m_{AB}\mathbf{v}_{AB} \tag{3.4.22}$$

In terms of these symbols, the canonical particle momenta (3.4.1) can be written as

$$\mathbf{p}_A = \frac{\partial \mathscr{L}}{\partial \mathbf{v}_A}$$

$$= \sum_{B}{}' m_{AB}\mathbf{v}_{AB}$$

$$= \sum_{B}{}' \mathbf{p}_{AB} \tag{3.4.23}$$

the Lagrangian (3.2.1) becomes

$$L(\mathbf{r}_{AB}, \mathbf{v}_{AB}) = \sum \sum_{A>B} \left(\frac{m_{AB}}{2}\ v_{AB}^2 + \phi_{AB}\right) \tag{3.4.24}$$

and the Hamiltonian (3.4.5) assumes the form

$$H(\mathbf{r}_{AB}, \mathbf{p}_{AB}) = E = \sum_{A>B}\sum \left(\frac{p_{AB}^2}{2m_{AB}} - \phi_{AB} \right)$$

$$= \sum_{A>B}\sum \frac{p_{AB}^2 r_{AB}}{4m_A m_B f\beta} - \sum_{A>B}\sum f \frac{m_A m_B}{r_{AB}} \qquad (3.4.25)$$

where we have set $c = 1$ in the last step. With the definition of $\mathbf{p}_A$ given as in (3.4.23), the Hamiltonian (3.4.25) now leads immediately to the canonical equations of motion:

$$\sum_A \dot{\mathbf{r}}_A \cdot d\mathbf{p}_A = \sum_A \mathbf{v}_A \cdot d\mathbf{p}_A$$

$$= \sum_A \mathbf{v}_A \cdot \sum_B{}' d\mathbf{p}_{AB}$$

$$= \sum_{A>B}\sum \frac{\mathbf{p}_{AB}}{m_{AB}} \cdot d\mathbf{p}_{AB} \qquad (3.4.26)$$

and

$$-\dot{\mathbf{p}}_A = \frac{\partial H}{\partial \mathbf{r}_A}$$

$$= \sum_B{}' \frac{\partial H}{\partial \mathbf{r}_{AB}}$$

$$= \sum_B{}' \left(\frac{\mathbf{r}_{AB}}{r_{AB}^3} fm_A m_B + \frac{\mathbf{r}_{AB}}{r_{AB}} \frac{p_{AB}^2}{4m_A m_B f\beta} \right)$$

$$= fm_A \sum_B{}' \frac{m_B}{r_{AB}^3} \mathbf{r}_{AB}(1 + \beta \mathbf{v}_{AB}^2) \qquad (3.4.27)$$

Comparison of equation (3.4.27) with equation (3.4.12) reveals that

$\dot{\mathbf{p}}_A$ in an N-particle cosmos is additively composed of independent two-particle contributions $\dot{\mathbf{p}}_{AB}$ from all particle pairs AB ($B \neq A$).

From the explicit expression (3.4.25) for the Hamiltonian $H(\mathbf{r}_A, \mathbf{p}_A) = H(\mathbf{r}_{AB}, \mathbf{p}_{AB})$, we have

$$\frac{\partial H}{\partial \mathbf{p}_{AB}} = \frac{\mathbf{p}_{AB}}{m_{AB}}$$

$$= \mathbf{v}_{AB} \tag{3.4.28}$$

Therefore, the familiar canonical equations (cf. Yourgrau and Mandelstam, 1960)

$$\frac{\partial H}{\partial \mathbf{p}_A} = \mathbf{v}_A \tag{3.4.29}$$

are valid if the connection between any canonical momentum $\mathbf{p}_A$ and the relative velocities $\mathbf{v}_{AB}$ is given by the formula (3.4.23):

$$\mathbf{p}_A = \sum_B' m_{AB} \mathbf{v}_{AB} \tag{3.4.30}$$

with

$$m_{AB} = m_{BA}$$

These equations, for all A, are the analytical expression of Mach's principle. According to (3.4.21) and (3.4.30), Mach's principle is a Hertzian anholonomic constraint existing among the particles in the universe: For every particle A, the condition (3.4.3), i.e.,

$$\mathbf{p}_A = -\sum_{B \neq A} \mathbf{p}_B$$

$$= -\sum_{B \neq A} \left(\frac{2\beta f}{c^2} m_B \sum_{D \neq B} \frac{m_D}{r_{DB}} \mathbf{v}_{BD} \right) \tag{3.4.31}$$

must be fulfilled. In words: Mach's principle requires the canonical momentum of any one particle to exactly balance the total momentum of all other particles in the universe (Treder, 1976b).

For an isotropic universe, in particular, this principle has the consequence that the fundamental astronomical system, defined with respect to the metagalaxy, must be identical with the inertial system defined relative to the motions of celestial objects. It is interesting that this consequence of the Mach–Einstein doctrine accords with the fact that the inner-planet inertial system and the "fixed-star" coordinate system coincide with one another, with an accuracy of about 0.4 sec per century in their rates of rotation.

Appendix A

Formalism of
Four-Dimensional Spaces

A.1. Differentiable Manifolds

Let x^k, with $k = 0, 1, 2, 3$ denote four variables, called *coordinates*, each of which can independently assume real values between given limits, and let x stand for the *set* of four coordinates. If $\varphi(x)$ is a function of the coordinates, then we shall use the notation

$$\partial_k \varphi = \varphi_{,k} = \frac{\partial \varphi}{\partial x^k} \qquad \text{(A.1.1)}$$

for the first-order derivatives of φ with respect to the x^k and the symbols $\varphi_{,k,l}$, etc. for the derivatives of higher order. All the functions employed by us will be assumed to possess derivatives at least of the orders 1 and 2, exhibiting discontinuities only exceptionally, viz., across certain surfaces. Such functions are referred to as *piecewise functions of class* C^2.

Starting from the coordinates x, one can introduce new coordinates x' with the aid of four functions $\varphi'^i(x)$:

$$x'^i = \varphi'^i(x) \tag{A.1.2}$$

these functions we shall require to be piecewise of class C^2 and to correspond to a transformation with nonzero Jacobian. The totality of points labeled by x or x' constitutes a four-dimensional *differentiable manifold* wherein changes of coordinates are reversible. The dimensionality of the manifold is chosen, of course, with the view of applying our formalism to general relativity theory.

A set of four equations

$$F_i(x) = 0, \qquad i = 0, 1, 2, 3 \tag{A.1.3}$$

defines a *point* or *event*, at the intersection of four hypersurfaces, corresponding to a fixed value for each of the x^k. At any such point, the four lines defined by the intersection curves themselves of the hypersurfaces can be chosen as coordinate lines, along each of which only one of the coordinates changes, while the other three remain constant. In this manner it becomes possible to erect at the point x^k a *tetrad* of unit vectors $dx^k/|\, dx^k\,|$, tangent to the lines of varying x^0, x^1, x^2, and x^3, respectively. At a neighboring point $x^k + dx^k$ another tetrad of unit tangent vectors can be constructed, by the same procedure, and so we may continue until all points of the manifold have been covered. The tetrad field thus established serves to define a linear coordinate system within an infinitesimal neighborhood of every point of the four-dimensional manifold. Such a local system of coordinates we call *natural*.

A.2. Metric Space

Space is called *metric* if a *squared distance* (or *line element*) between any two infinitesimally close points belonging to it is defined by an invariant homogeneous quadratic function of the coordinate differentials,

$$ds^2 = g_{ik}\, dx^i\, dx^k \qquad \text{(A.2.1)}$$

where the *metric functions* $g_{ik}(x)$ are piecewise of the class C^2 and summation over repeated indices is implied.

We shall assume four-dimensional metric space, denoted by M^4, to be of the *normal hyperbolic type*, that is, of a structure such that, in the infinitesimal neighborhood of every point, a coordinate transformation $x \to x_o$ can be found which casts ds^2 in the pure square form

$$ds^2 = (dx_o^0)^2 - (dx_o^1)^2 - (dx_o^2)^2 - (dx_o^3)^2 \qquad \text{(A.2.2)}$$

This means that the symmetric parts of the metric coefficients, now written η_{ik}, assume the diagonal form

$$(\eta_{ik}) = \text{diag}(1, -1, -1, -1) \qquad \text{(A.2.3)}$$

We note that, ds^2 being an invariant under the transformation

$$dx^i = \frac{\partial x^i}{\partial x_o^k}\, dx_o^k \qquad \text{(A.2.4)}$$

the relation between the symmetric parts of the metrics g_{ik} and

η_{ik} is obviously given by

$$g_{ik} = \frac{\partial x_o^l}{\partial x^i} \frac{\partial x_o^m}{\partial x^k} \eta_{lm} \qquad \text{(A.2.5a)}$$

and, conversely,

$$\eta_{lm} = \frac{\partial x^i}{\partial x_o^l} \frac{\partial x^k}{\partial x_o^m} g_i \qquad \text{(A.2.5b)}$$

In view of equation (A.2.2), the natural coordinate system associated with the metric η_{ik} is orthogonal, and the infinitesimal neighborhood of any point is said to be of the *pseudo-Euclidean* type. The physical space thus defined is known as *space–time*. It belongs to the *Riemannian* variety, designated by R^4, if the symmetry condition

$$g_{ik} = g_{ki} \qquad \text{(A.2.6)}$$

is fulfilled. The locally pseudo-Euclidean regions of space–time, for which the designation E^4 is employed, can be viewed as "tangent" to R^4. The space E^4 is *Minkowskian* in nature, its structure being defined by the kinematical axioms of special relativity theory.

According to equation (A.2.2), one can set $ds^2 = 0$ without demanding that all the dx_o^i vanish. Hence, in every point of space–time, it is possible to define an *elementary* isotropic *cone* by $ds^2 = 0$. With reference to this cone, one can make an invariant classification of the infinitesimal vectors $d\mathbf{x} = (dx^0, dx^1, dx^2, dx^3)$ whose components dx^i are given in terms of the coordinate differentials dx_o^i by arbitrary transformations of the type (A.2.4). We call a vector

$d\mathbf{x}$ *timelike*, *isotropic*, or *spacelike* depending on whether

$$ds^2 = \eta_{ik}\, dx_o^i\, dx_o^k = g_{ik}\, dx^i\, dx^k$$

is positive, zero, or negative.

One defines a *line* Γ by a quartet of functions

$$x^k = \varphi^k(\sigma), \qquad k = 0, 1, 2, 3 \tag{A.2.7}$$

of some continuous parameter σ, which may be s itself or some function of it. At any point of Γ where a directed line element $d\mathbf{x}$ is timelike, Γ is timelike, too; and corresponding statements hold for the isotropic and spacelike orientations. From a purely mathematical viewpoint, the orientation can vary rather arbitrarily along one and the same line. But if the line is to have a physical meaning, then the question may arise whether such a variation of its orientation from point to point makes physical sense.

A.3. Contravariance and Covariance

A set of four vectors $\mathbf{e}_i$ constitutes a *basis* if it is possible to write the vector separation $d\mathbf{x}$ between neighboring points of space–time as the sum

$$d\mathbf{x} = \mathbf{e}_i\, dx^i \tag{A.3.1}$$

Consider now the coordinate transformation $x \to x'$ and its reverse, which are defined by the equations

$$dx'^i = a'^i_k \, dx^k, \qquad a'^i_k = \frac{\partial x'^i}{\partial x^k} \qquad (A.3.2a)$$

$$dx^k = a^k_i \, dx'^i, \qquad a^k_i = \frac{\partial x^k}{\partial x'^i} \qquad (A.3.2b)$$

This change of coordinates leads of course to a new basis $\mathbf{e}_i'$, such that

$$\mathbf{e}_i' \, dx'^i = \mathbf{e}_k \, dx^k \qquad (A.3.3)$$

Substitution from (A.3.2) and (A.3.3) then gives

$$\mathbf{e}_i' \, dx'^i = \mathbf{e}_k a^k_i \, dx'^i$$

and

$$\mathbf{e}_i' a'^i_k \, dx^k = \mathbf{e}_k \, dx^k$$

Consequently, one sees that

$$\mathbf{e}_i' = \mathbf{e}_k a^k_i \qquad (A.3.4a)$$

and

$$\mathbf{e}_k = \mathbf{e}_i' a'^i_k \qquad (A.3.4b)$$

Equations (A.3.2) and (A.3.4) enable us to define the properties of contravariance and covariance, respectively. Quite generally, a set of quantities $A^{\alpha\beta\cdots}_{\lambda\mu\cdots}$ with $\alpha, \lambda, \ldots = 0, 1, 2, 3$, are said to define a tensor that is *contravariant* with respect to its superscripts $\alpha, \beta, \ldots$ and *covariant* with respect to its subscripts $\lambda, \mu, \ldots$ if

Formalism of Four-Dimensional Spaces

they transform as follows under coordinate changes $x \to x'$:

$$A'^{ij\cdots}_{pq\cdots} = a'^i_{\ \alpha} a'^j_{\ \beta} \cdots a^\lambda_{\ p} a^\mu_{\ q} \cdots A^{\alpha\beta\cdots}_{\lambda\mu\cdots} \qquad (A.3.5)$$

In other words, tensor components transform like the products of coordinate differentials in regard to their contravariant indices and like basis vectors relative to their covariant indices. We shall follow the common practice of denoting both the tensor and its components by the same symbol $A^{ij\cdots}_{pq\cdots}$ when no possibility of confusion exists.

The *rank* of a tensor is defined by the number of its upper and lower indices combined. The simplest tensors are of rank 0; known as *invariants* or *scalars*, they are of special importance, since their values do not change from one coordinate system to another. [Some authors distinguish between (mathematical) invariants and (physical) scalars.] Tensors of rank 1 are commonplace, appearing either as *contravariant vectors* V^i or *covariant vectors* V_i. The more complicated tensors of second and higher ranks, such as T^{ijk} or T_{ijk}, can exhibit a purely contravariant or covariant character, but generally they are of a type such as $T^{ij}_{\ \ k}$, with mixed transformation properties. In metric space, the pivotal invariant is, of course, the squared distance (A.2.1) between neighboring points. Its postulated invariance under the transformation $x \to x'$,

$$g'_{\alpha\beta}\, dx'^\alpha\, dx'^\beta = g_{\lambda\mu}\, dx^\lambda\, dx^\mu$$

$$= g_{\lambda\mu} a^\lambda_{\ \alpha} a^\mu_{\ \beta}\, dx'^\alpha\, dx'^\beta$$

leads to the transformation rule

$$g'_{\alpha\beta} = a^\lambda_{\ \alpha} a^\mu_{\ \beta} g_{\lambda\mu} \qquad (A.3.6)$$

for the metric coefficients. Accordingly, $g_{\alpha\beta}$ is seen to be a covariant tensor of rank 2, whence its usual designation as the *metric tensor*.

A.4. The Affine Connection

It is possible to define the so-called *affine connection* independently of the existence of a metric. For this purpose, consider a manifold of points x, $x + dx$, etc. and let $\mathbf{e}_k$ define a basis at the point x such that $d\mathbf{x} = \mathbf{e}_i \, dx^i$; in the point $x + dx$ the basis will then be $\mathbf{e}_k + d\mathbf{e}_k$; etc. If we now perform the coordinate transformation $x \to x'$, the question arises: What is the form of the coordinate transformation at the point $x + dx$ if its form is known at the point x?

Well, it should be possible to expand $d\mathbf{e}_k$ in terms of the basis vectors $\mathbf{e}_i$; and let it furthermore be assumed that the expansion coefficients are linear combinations of the dx^i. Accordingly, we define the quantities Γ^i_{jk} such that

$$d\mathbf{e}_j = \Gamma^i_{jk}\mathbf{e}_i \, dx^k \qquad \text{(A.4.1)}$$

whereby we have achieved that $d\mathbf{e}_j$, like $d\mathbf{x}$, is written as a linear form of the dx^k. Use of the Γ^i_{jk} thus allows us to *connect* the basis vectors $\mathbf{e}_k$ at neighboring points throughout any four-dimensional space; and so-called *affine space* can be constructed point by point if the Γ^i_{jk} are prescribed as functions of the x^i. For this reason, the quantities Γ^i_{jk} are called the *components* of the *affine connection*. Despite its appearance, Γ^i_{jk} is not a tensor, as will become evident further on.

A.5. Integrability, Torsion, and Curvature

Returning to metric space M^4, defined by $g_{ik}(x)$, we postulate the symmetry of the tensor g_{ik}.

Next we introduce a natural coordinate system in the point x of M^4, but extend its axes while holding fixed the values of the coefficients g_{ik}, which now are denoted by η_{ik}. The resulting pseudo-Cartesian axes give rise to coordinates $x_o{}^i$ and a squared distance that is everywhere

$$d\sigma^2 = \eta_{ik}\, dx_o^i\, dx_o^k \qquad\qquad \text{(A.5.1)}$$

In this quadratic form, which is the defining property of pseudo-Euclidean space M_o^4, the coefficients can be rewritten with the aid of the constant basis $(\mathbf{e}_i)_o$, equal to $\mathbf{e}_i$ in M^4, as

$$\eta_{ik} = (\mathbf{e}_i)_o \cdot (\mathbf{e}_k)_o \qquad\qquad \text{(A.5.2)}$$

The foregoing procedure can be repeated at each point of M^4. And it should be noted that, up to first-order magnitudes, relations that are valid in M^4 also hold in M_o^4; thus

$$d\mathbf{x} = (d\mathbf{x})_o \qquad \text{and} \qquad \left(\frac{\partial \mathbf{x}}{\partial x_o^k}\right)_o = \frac{\partial \mathbf{x}}{\partial x^k}$$

However, a difference does exist in that

$$(d\mathbf{e}_k)_o \neq d(\mathbf{e}_k)_o \qquad\qquad \text{(A.5.3)}$$

Appendix A

This statement follows from the fact that

$$(d\mathbf{e}_k)_o = (\Gamma^i_{kl})_o(\mathbf{e}_i)_o(dx^l)_o$$
$$= (\Gamma^i_{kl})_o(\mathbf{e}_i)_o \, dx^l_o$$
$$= \Gamma^i_{kl}\mathbf{e}_i \, dx^l_o$$

while

$$d(\mathbf{e}_k)_o = \Gamma^i_{kl}\mathbf{e}_i \, dx^l \qquad \text{with } dx^l \neq dx^l_o$$

If one should postulate the equality of $(d\mathbf{e}_k)_o$ and $d(\mathbf{e}_k)_o$, it would mean that

$$(\mathbf{e}_i)_o \cdot (d\mathbf{e}_k)_o + (\mathbf{e}_k)_o \cdot (d\mathbf{e}_i)_o = (\mathbf{e}_i)_o \cdot d(\mathbf{e}_k)_o + (\mathbf{e}_k)_o \cdot d(\mathbf{e}_i)_o$$

and, therefore,

$$(dg_{ik})_o = [d(\mathbf{e}_i \cdot \mathbf{e}_k)]_o = d(g_{ik})_o$$

Otherwise, we have

$$\mathbf{e}_i = (\mathbf{e}_i)_o + (\Gamma^k_{il})_o(x^l - x^l_o)(\mathbf{e}_k)_o$$

and

$$(d\mathbf{x})_o = (\mathbf{e}_i \, dx^i)_o$$
$$= [(\mathbf{e}_i)_o \, dx^i + (\Gamma^k_{il})_o(x^l - x^l_o) \, dx^i(\mathbf{e}_k)_o]_o$$
$$= (\mathbf{e}_k)_o[dx^k + (\Gamma^k_{il})_o(x^l - x^l_o) \, dx^i]_o$$

This means, if $i \neq l$, that

$$\left(\frac{\partial \mathbf{x}}{\partial x^i}\right)_o = (\mathbf{e}_i)_o + (\mathbf{e}_k)_o (\Gamma^k_{il})_o (x^l - x^l_o)$$

and, therefore,

$$\left(\frac{\partial^2 \mathbf{x}}{\partial x^l \, \partial x^i}\right)_o = (\mathbf{e}_k)_o (\Gamma^k_{il})_o$$

It consequently follows in general that

$$\left(\frac{\partial^2 \mathbf{x}}{\partial x^l \, \partial x^i}\right)_o \neq \left(\frac{\partial^2 \mathbf{x}}{\partial x^i \, \partial x^l}\right)_o$$

i.e., the integral of $d\mathbf{x}$ along a closed path will not vanish in M^4_o unless

$$\Gamma^k_{il} = \Gamma^k_{li} \tag{A.5.4}$$

The best way to define a *Riemannian space* R^4 is by assuming the integrability of lines in M^4. This means, according to (A.5.4), that a necessary and sufficient condition for the Riemannian character of space is the symmetry of its affine connection Γ^i_{kl} relative to its subscripts.

In M^4 a *torsion* can be defined by the contravariant vector

$$\Omega^i = \tfrac{1}{2}(\Gamma^i_{jk} - \Gamma^i_{kj})(dx^k \, \delta x^j - dx^j \, \delta x^k) \tag{A.5.5}$$

where $d\mathbf{x}$ and $\delta\mathbf{x}$ are taken along the adjacent sides of an infinitesimal parallelogram whose area has the projections

$$\tfrac{1}{2}(dx^k \, \delta x^j - dx^j \, \delta x^k)$$

Consequently, metric spaces in general possess a nonzero torsion, whereas the torsion vanishes in Riemannian space.

One can further define in M^4 a *curvature* by the tensor

$$\Omega_k^l = \tfrac{1}{2} R^l{}_{kpq}(dx^q \, \delta x^p - dx^p \, \delta x^q) \qquad (A.5.6)$$

where $R^l{}_{kpq}$, called the *Riemann–Christoffel tensor*, has the definition

$$R^l{}_{kpq} = \Gamma^l{}_{kp,q} - \Gamma^l{}_{kq,p} + \Gamma^s{}_{kp}\Gamma^l{}_{sq} - \Gamma^s{}_{kq}\Gamma^l{}_{sp} \qquad (A.5.7)$$

with the comma denoting partial differentiation of the kind

$$\Gamma^c{}_{ab,j} = \frac{\partial \Gamma^c{}_{ab}}{\partial x^j} \qquad (A.5.8)$$

In the tangent space M_o^4, where the g_{ik} are constant, the components Γ^i_{jk} of the affine connection clearly vanish, since the $\mathbf{e}_i$ do not change with position. The tangent space therefore possesses neither curvature nor torsion; it is Euclidean or, rather, pseudo-Euclidean in view of the alternation of signs of the metric $g_{ik} = \eta_{ik}$. In M_o^4 one can have no objection to writing $g_{ik} = g_{ki}$, because the quadratic form $ds^2 = g_{ik} \, dx^i \, dx^k$ determines only the symmetric combinations $g_{ik} + g_{ki}$, not the g_{ik} $(i \neq k)$ separately. Assuming the symmetry of the metric tensor therefore does not imply any restriction.

Conversely, it can be proven: (a) if $R^l{}_{kpq} = 0$, everywhere and for all indices, then space is pseudo-Euclidean or *flat*; (b) if $R^l{}_{kpq} \neq 0$ but $\Omega^i = 0$, then space is Riemannian.

A.6. Transition from Covariance to Contravariance, and Conversely, in Metric Space

Suppose the four vectors $\mathbf{e}_i$ provide a basis in M^4. Then $d\mathbf{x} = \mathbf{e}_i \, dx^i$ and, hence,

$$ds^2 = d\mathbf{x} \cdot d\mathbf{x}$$
$$= \mathbf{e}_i \cdot \mathbf{e}_i \, dx^i \, dx^k \qquad (A.6.1)$$

Thus, in a metric space,

$$g_{ik} = \mathbf{e}_i \cdot \mathbf{e}_k \qquad (A.6.2)$$

If we denote by

$$g = \det g_{ik} \qquad (A.6.3)$$

the determinant of these quantities and by γ^{ik} their cofactors, then the well-known development

$$g_{ik}\gamma^{ik} = g \qquad (A.6.4)$$

obtains. As a result, the quantities

$$g^{ik} = g^{-1}\gamma^{ik} \qquad (A.6.5)$$

define a contravariant tensor of rank 2, since the double inner product with the covariant tensor g_{ik} gives

$$g_{ik}g^{ik} = 1 \qquad (A.6.6)$$

an invariant, which is possible only if g^{ik} itself is a contravariant tensor.

From the familiar properties of determinants, one also knows that

$$g_{ij}g^{jk} = \delta_i^k \qquad (A.6.7)$$

where $\delta_i^{\ k}$ stands for the *Kronecker symbol*. Equation (A.6.7) demonstrates, first, that the matrix with elements g^{ik} is the inverse of that with the elements g_{ik} and, secondly, that the Kronecker symbol is indeed a mixed tensor (the same in all coordinate systems), as the notation δ_i^k suggests.

The tensors g_{ik} and g^{ik} play very useful mathematical roles in tensor algebra. To see this, let

$$\mathbf{A} = A^k \mathbf{e}_k \qquad (A.6.8)$$

be a vector with *contravariant* components A^k. The *covariant* components of $\mathbf{A}$ will then, per definition, be

$$A_k = \mathbf{A} \cdot \mathbf{e}_k = A^i \mathbf{e}_i \cdot \mathbf{e}_k = A^i g_{ik} = g_{ki} A^i \qquad (A.6.9)$$

Thus g_{ik} converts contravariant vector components into covariant ones. The converse operation is effected by g^{ik}, since we have

$$g^{ik} A_k = g^{ik} A^j g_{jk} = \delta_i^j A^j = A^i \qquad (A.6.10)$$

where use has been made of equation (A.6.7) in the second step.

This ability of the tensors g_{ik} and g^{ik} to, respectively, *lower* and *raise* indices is not confined to vectors; the procedure is equally valid for tensors of any rank. So, for example, it is legitimate to set

$$g_{pk}A^{ijk} = A^{ij}{}_p$$

and

$$g^{ir}A_{pqr} = A_{pq}{}^i$$

since the left-hand sides have indeed the transformation properties suggested by the symbols on the right. It should be noted that the lowering or raising of any index leaves its horizontal position unchanged.

A.7. Relation between the Affine and Metric Connections in R^4

We recall that, per definition of the affine connection,

$$d\mathbf{e}_k = \Gamma^i_{kl}\mathbf{e}_i \, dx^l \tag{A.7.1}$$

and, therefore,

$$
\begin{aligned}
dg_{jk} &= d(\mathbf{e}_j \cdot \mathbf{e}_k) \\
&= \mathbf{e}_j \cdot \mathbf{e}_i \Gamma^i_{kl} \, dx^l + \mathbf{e}_i \cdot \mathbf{e}_k \Gamma^i_{jl} \, dx^l \\
&= (g_{ij}\Gamma^i_{kl} + g_{ik}\Gamma^i_{jl}) \, dx^l
\end{aligned}
$$

It follows that

$$g_{jk,l} = g_{ij}\Gamma^i_{kl} + g_{ik}\Gamma^i_{j} \tag{A.7.2}$$

By repeated application of this formula, one finds that

$$[kl, m] = \tfrac{1}{2}(g_{km,l} + g_{lm,k} - g_{kl,m})$$
$$= \Gamma^i_{kl} g_{im} \qquad (A.7.3)$$

The function $[kl, m]$ defined here is called the *Christoffel symbol of the first kind*.

Multiplication of the foregoing equation by g^{im} and application of equation (A.6.7) immediately lead to

$$\begin{Bmatrix} i \\ kl \end{Bmatrix} = \tfrac{1}{2} g^{im}(g_{km,l} + g_{lm,k} - g_{kl,m})$$
$$= \Gamma^i_{kl} \qquad (A.7.4)$$

The expression

$$\begin{Bmatrix} i \\ kl \end{Bmatrix} = g^{im}[kl, m] \qquad (A.7.5)$$

is commonly referred to as the *Christoffel symbol of the second kind*. In R^4 it is obviously synonymous with the affine connection, and accordingly we shall henceforth write equation (A.7.1) in R^4 in the form

$$d\mathbf{e}_k = \begin{Bmatrix} i \\ kl \end{Bmatrix} \mathbf{e}_i \, dx^l \qquad (A.7.6)$$

A.8. Absolute Differential and Covariant Differentiation

Let us consider a vector field

$$\mathbf{A}(x) = A^i \mathbf{e}_i \qquad (A.8.1)$$

in an affine space A^4. The change in $\mathbf{A}(x)$ resulting from arbitrary displacements dx^i is

$$d\mathbf{A} = d\mathbf{A}^i\mathbf{e}_i + A^k\,d\mathbf{e}_k$$
$$= (dA^i + A^k\Gamma^i_{kl}\,dx^l)\mathbf{e}_i$$

which can be written as

$$d\mathbf{A} = DA^i\mathbf{e}_i \qquad\qquad (A.8.2)$$

if we define

$$DA^i = dA^i + \Gamma^i_{kl}A^k\,dx^l \qquad\qquad (A.8.3)$$

DA^i is called the *absolute differential* of A^i; by virtue of equation (A.8.2), it is a contravariant vector.

The absolute differential DA^i clearly is the sum of two terms: The first one expresses simply the increase in the component A^i when one passes from point x to point $x + dx$; the second is the contribution to DA^i stemming from the modification of the natural coordinate system during the same passage.

We are thus led to define the *covariant derivative of a contravariant vector* as the second-rank tensor

$$A^i_{;k} \equiv \nabla_k A^i = \frac{DA^i}{dx^k} = A^i_{,k} + \Gamma^i_{jk}A^j \qquad\qquad (A.8.4)$$

which one can denote by either $A^i_{;k}$ or $\nabla_k A^i$.

The contravariant vector A^i has a covariant counterpart

$$A_i = g_{ij}A^j = \mathbf{e}_i \cdot \mathbf{e}_j A^j = \mathbf{A} \cdot \mathbf{e}_i$$

165

such that $A^i A_i$ is an invariant. The associated increments $d\mathbf{A} \cdot \mathbf{e}_i$, which we write as DA_i, are given by

$$DA_i = d\mathbf{A} \cdot \mathbf{e}_i$$

$$= d(\mathbf{A} \cdot \mathbf{e}_i) - \mathbf{A} \cdot d\mathbf{e}_i$$

$$= dA_i - \mathbf{A} \cdot \Gamma^k_{il}\mathbf{e}_k \, dx^l$$

Therefore,

$$DA_i = dA_i - \Gamma^k_{il}A_k \, dx^l \tag{A.8.5}$$

We thus arrive at the following definition of the *covariant derivative of a covariant vector*:

$$A_{i;l} \equiv \nabla_l A_i$$

$$= A_{i,l} - \Gamma^k_{il}A_k \tag{A.8.6}$$

For a mixed tensor of arbitrary rank, covariant differentiation adds an affine term to the ordinary derivative for each contravariant superscript and subtracts a similar term for each covariant subscript. Thus, for example,

$$A_{ik}{}^{pq}{}_{;l} \equiv \nabla_l A_{ik}{}^{pq}$$

$$= A_{ik}{}^{pq}{}_{,l} + \Gamma^p_{sl}A_{ik}{}^{sq} + \Gamma^q_{sl}A_{ik}{}^{ps} - \Gamma^s_{il}A_{sk}{}^{pq} - \Gamma^s_{kl}A_{is}{}^{pq} \tag{A.8.7}$$

In a *metric* space, Γ^i_{jk} is of course the same as $\{^i_{jk}\}$. It can furthermore be verified that

$$DA_i = g_{ik}DA^k$$

In a *Riemannian* space, it is particularly interesting to note that the covariant derivative of the metric tensor vanishes identically:

$$g_{jk;l} = g_{jk,l} - \Gamma_{jl}^i g_{ik} - \Gamma_{kl}^i g_{ji} = 0 \qquad (A.8.8)$$

which follows from equation (A.7.2), since, in R^4, $g_{ji} = g_{ij}$.

A.9. Parallel Displacement or Transport

We are now able to assign a meaning to the statement that two vectors at *neighboring* points of curved space are parallel to one another. We found above that

$$dA^k = DA^k - \Gamma_{lm}^k A^l \, dx^m \qquad (A.9.1a)$$

and

$$dA_k = DA_k + \Gamma_{km}^l A_l \, dx^m \qquad (A.9.1b)$$

which means that the differential of **A** corresponding to an arbitrary displacement exceeds its absolute differential; the extra term, which we designate by

$$(dA^k)_{\parallel} = -\Gamma_{lm}^k A^l \, dx^m \qquad (A.9.2a)$$

and

$$(dA_k)_{\parallel} = \Gamma_{km}^l A_l \, dx^m \qquad (A.9.2b)$$

respectively, defines the change in the vector **A** that occurs when it is transported between the infinitesimally close points x and $x + dx$ by the so-called procedure of *parallel displacement*.

167

In R^4, the *length* of a vector is conserved during a parallel displacement. Indeed, the differential of the squared length,

$$\mathbf{A} \cdot \mathbf{A} = g_{ij}A^iA^j \tag{A.9.3}$$

is furnished by

$$d(\mathbf{A} \cdot \mathbf{A}) = dg_{ij}A^iA^j + 2g_{ij}A^i\, dA^j$$

and, therefore,

$$
\begin{aligned}
[d(\mathbf{A} \cdot \mathbf{A})]_{\shortparallel} &= (dg_{ij})_{\shortparallel}A^iA^j - 2g_{ij}A^i\Gamma^j_{km}A^k\, dx^m \\
&= (dg_{ik} - 2g_{ij}\Gamma^j_{km}\, dx^m)A^iA^k \\
&= (g_{ik,m} - g_{ij}\Gamma^j_{km} - g_{kj}\Gamma^j_{im})\, dx^m A^iA^k \\
&= g_{ik;m}\, dx^m A^iA^k
\end{aligned}
$$

If we recall that $g_{ik;m} = 0$, the desired result

$$[d(\mathbf{A} \cdot \mathbf{A})]_{\shortparallel} = 0 \tag{A.9.4}$$

follows.

A.10. Symmetry and Other Properties of Torsion and Various Curvatures

We apply the name *torsion* to the contravariant vector

$$
\begin{aligned}
\Omega^i &= \tfrac{1}{2}(\Gamma^i_{jk} - \Gamma^i_{kj})(dx^k\, \delta x^j - dx^j\, \delta x^k) \\
&\equiv \tau^i_{jk}(-d^2\sigma^{jk})
\end{aligned}
\tag{A.10.1}
$$

or, alternatively, to the quantity

$$\tau^i_{jk} = \Gamma^i_{jk} - \Gamma^i_{kj} = -\tau^i_{kj} \qquad \text{(A.10.2)}$$

The tensor τ^i_{jk} and the projection

$$d^2\sigma^{jk} = \tfrac{1}{2}(dx^j \delta x^k - dx^k \delta x^j)$$
$$= -d^2\sigma^{kj} \qquad \text{(A.10.3)}$$

are antisymmetric in their lower and upper indices, respectively. Because of the symmetry of the affine connection in R^4 with regard to its subscripts, $\tau^i_{jk} = 0$ in Riemannian space.

We recall that the curvature of metric space is determined by the Riemann–Christoffel tensor (A.5.7), which has the antisymmetry property

$$R^i{}_{kpq} = -R^i{}_{kpq} \qquad \text{(A.10.4)}$$

An appropriate summation over the antisymmetric indices of this tensor produces the *curvature of rotation,*

$$\Omega^i_k = R^i_{kpq}(-d^2\sigma^{pq}) \qquad \text{(A.10.5)}$$

Contraction of the tensor Ω^i_k, which involves putting $i = k$ and summing over i, finally yields the scalar

$$\Omega = \Omega^k_k = R^k_{kpq}(-d^2\sigma^{pq}) \qquad \text{(A.10.6)}$$

known as the *homothetic curvature.*

169

The foregoing considerations suggest the classification of metrics that now follows.

A.11. Classification of Metric Types

A.11.1. Classification from the Integrability Point of View

There are three classes of metric spaces:

(a) E^4 (Euclidean):

$$\oint d\mathbf{x} = 0, \qquad \oint d\mathbf{e}_i = 0 \qquad \text{(A.11.1a)}$$

neither torsion, nor curvature;

$$g_{ik} = g_{ki} = \text{const} \qquad \text{(A.11.1b)}$$

$$\Gamma^i_{jk} = 0 \qquad \text{(A.11.1c)}$$

(b) R^4 (Riemannian):

$$\oint d\mathbf{x} = 0, \qquad \oint d\mathbf{e}_i \neq 0 \qquad \text{(A.11.2a)}$$

nonzero curvature, but no torsion;

$$g_{ik}(x) = g_{ki}(x) \qquad \text{(A.11.2b)}$$

$$\Gamma^i_{kj} = \Gamma^i_{jk} = \left\{ \begin{matrix} i \\ jk \end{matrix} \right\} \qquad \text{(A.11.2c)}$$

(c) M^4 (other):

$$\oint d\mathbf{x} \neq 0, \qquad \oint d\mathbf{e}_i \neq 0 \qquad \text{(A.11.3a)}$$

nonzero torsion and curvature;

$$\Gamma^i_{kj} \neq \Gamma^i_{jk} \neq \left\{ \begin{matrix} i \\ jk \end{matrix} \right\} \qquad \text{(A.11.3b)}$$

A.11.2. Classification from the Affine Connection

Having shown that the affine connection Γ^i_{jk} is identical with the Christoffel symbol $\{^i_{jk}\}$ in a Riemannian space, we shall henceforth employ the notation A^i_{jk} for the components of an affine connection which do not necessarily pertain to a Riemannian space. The only remaining required property of a general differentiable manifold V_n is then the existence of an affine connection. The latter, it should be emphasized, is not a tensor, since its transformation rule in V_n can be proved to read as follows:

$$A'^i_{kl} = A^d_{ef} a'^i{}_d a^e{}_k a^f{}_l + \frac{\partial a^d_l}{\partial x'^k} a'^i{}_d \qquad \text{(A.11.4)}$$

for a coordinate transformation $x \to x'$. Let us designate the affine components appearing in equation (A.11.4) by $A^i_{kl}{}_1$ and $A'^i_{kl}{}_1$; then this equation is seen to remain unaltered if the A's are replaced by

$$A^i_{kl}{}_2 = A^i_{kl}{}_1 + \Omega^i_{kl} \qquad \text{(A.11.5)}$$

and similarly for the primed A's, where Ω^i_{kl} is a tensor of rank 3, which of course transforms according to

$$\Omega'^{i}_{kl} = a'^{i}_{d}a^{e}_{k}a^{f}_{l}\Omega^{d}_{ef} \tag{A.11.6}$$

In the space V_n, a *covariant derivative*, e.g., of a vector B^i will be defined by

$$B^{i}_{;k} = B^{i}_{,k} + A^{i}_{jk}B^{j} \tag{A.11.7}$$

If a metric exists in V_n, its covariant derivative will be, as in equation (A.8.8),

$$g_{ik;l} = g_{ik,l} - A_{ikl} - A_{kil} \tag{A.11.8}$$

with the definition

$$A_{ikl} = g_{im}A^{m}_{kl} \tag{A.11.9}$$

In case

$$g_{ik;l} = 0 \tag{A.11.10}$$

that is, if the metric is covariantly constant with respect to the affine connection, then this connection itself and the transport it allows will be said to be *metrical*. In the *nonmetrical* situation,

$$g_{ik;l} \equiv \varphi_{ikl} \neq 0 \tag{A.11.11}$$

where, in view of the symmetry of the g_{ik}, the "inhomogeneity"

φ_{ikl} is necessarily symmetric in its first two indices:

$$\varphi_{ikl} = \varphi_{kil} \qquad \text{(A.11.12)}$$

Equation (A.11.10) is the mathematical expression of *Ricci's lemma*. One observes that, while there are n^3 coefficients A_{ikl}, only

$$[\tfrac{1}{2}n(n-1) + n]n = \tfrac{1}{2}n^2(n+1)$$

equations occur in Ricci's lemma; n denotes the dimensionality of space, which for space–time is 4. Consequently, Ricci's lemma does not allow us to determine all the coefficients of a metrical affine connection; in fact, $\tfrac{1}{2}n^2(n-1)$ quantities remain undetermined and, from (A.11.5), they form a tensor of rank 3.

If the A_{ikl} are symmetric in their last two indices,

$$A_{ikl} = A_{ilk} \qquad \text{(A.11.13)}$$

then they become identical with the Christoffel symbols

$$A^i_{kl} = g^{im}\Gamma_{mkl} \qquad \text{(A.11.14)}$$

But then all the coefficients

$$\underset{2}{A}_{ikl} \equiv \Gamma_{ikl} + \beta_{ikl}$$

are solutions of Ricci's lemma (A.11.10), as long as the third-rank tensor β_{ikl} is antisymmetric in its first two indices,

$$\beta_{ikl} = -\beta_{kil} \qquad \text{(A.11.15)}$$

Equation (A.11.15) is therefore a necessary and sufficient condition for the fulfillment of Ricci's lemma, i.e., for defining a metrical connection. The β_{ikl} can be regarded as *generalized Ricci coefficients*. Accordingly, the coefficients of the metrical connection belonging to a metric g_{ik} are additively composed of the corresponding Christoffel symbols and the generalized Ricci coefficients.

Since the β_{ikl} are antisymmetric in their first two indices, they cannot, insofar as they are nonzero, be symmetric in their last two indices:

$$\beta_{ikl} \neq \beta_{ilk} \qquad (A.11.16)$$

They can furthermore be expressed as linear combinations of the torsion components τ_{ikl}:

$$\Omega_{ikl} = \beta_{ikl}$$
$$= -\tau_{kil} - \tau_{lik} + \tau_{ikl} \qquad (A.11.17)$$

where

$$\tau_{ikl} = -\tau_{ilk}$$

Conversely, one finds that

$$\tau_{ikl} = -\beta_{ilk} + \beta_{lki} + \beta_{ikl} \qquad (A.11.18)$$

As a result, every solution of Ricci's lemma that differs from the Christoffel symbols describes a connection that does not fulfill the closure condition along an infinitesimal parallelogram.

If, by contrast, the Christoffel symbols are combined additively with a tensor ϱ_{ikl} that is symmetric in its first two indices,

$$A_{ikl} = \Gamma_{ikl} + \varrho_{ikl}$$

$$\varrho_{ikl} = \varrho_{kil}$$

(A.11.19)

then Ricci's lemma is not satisfied. Instead

$$g_{ik;l} = -\varrho_{ikl} - \varrho_{kil}$$

$$= -2\varrho_{ikl}$$

$$\equiv \varphi_{ikl}$$

(A.11.20)

and, in the most general case, equation (A.11.5) can be cast in the explicit form

$$A_{ikl} = \Gamma_{ikl} + \Omega_{ikl}$$

$$= \Gamma_{ikl} + \beta_{ikl} + \varrho_{ikl}$$

(A.11.21)

here Γ_{ikl} is the Christoffel symbol referring to the metric g_{ik}, β_{ikl} is the concomittant of the torsion, and $-2\varrho_{ikl}$ represents the inhomogeneity in the covariant derivative of the metric.

The expression (A.11.21) allows a classification of the geometries of a space V_n possessing a metric g_{ik}. This classification rests on a consideration of vector displacements, for which a straightforward calculation, based on equation (A.11.7) and (A.11.9), would show that, given a line $x^l = x^l(\tau)$ in terms of a parameter τ and a vector B^i that is displaced along this line, then the following equalities are satisfied:

$$(B^i B^k g_{ik})_{;l} \frac{dx^l}{d\tau} = (B^i B_i)_{;l} \frac{dx^l}{d\tau}$$

$$= (B^i B_i)_{,l} \frac{dx^l}{d\tau} \qquad \text{(A.11.22)}$$

We may distinguish three important cases:

(i) The simplest assumption yields the so-called Riemann–Christoffel or *infinitesimal parallel displacement*. It obtains when

$$\varrho_{ikl} = 0$$

$$\beta_{ikl} = 0 \qquad \text{(A.11.23)}$$

At the same time, the torsion vanishes, for

$$A_{ikl} - A_{ilk} = 0 \qquad \text{(A.11.24)}$$

The displacement is, however, not parallel over long distances, because the Riemann–Christoffel curvature tensor $R^i{}_{klm}$, with the definition (A.5.7), does not vanish.

Nevertheless, parallel displacement allows, according to Levi–Civita, a distant comparison between scalars and, consequently, between proper magnitudes, in the physical sense of self-measurable magnitudes (e.g., vector lengths). Indeed, if $x^l = x^l(\tau)$ defines a world-line, in terms of a parameter τ, and a vector B^i is displaced along this line, then the validity of Ricci's lemma enables us to write

$$B^i B^k g_{ik;l} \frac{dx^l}{d\tau} = 0 \qquad \text{(A.11.25)}$$

and

$$B^i{}_{;l} \frac{dx^l}{d\tau} = 0 \qquad (A.11.26)$$

In the present case, the Riemann–Christoffel tensor possesses all the Ricci symmetries

$$R_{ikmn} = -R_{iknm} = -R_{kimn} = R_{mnik} \qquad (A.11.27)$$

and

$$R_{iklm} + R_{ilmk} + R_{imkl} = 0 \qquad (A.11.28)$$

and in general it does not vanish.

(ii) The displacement is *simply metrical* if, in equation (A.11.21),

$$\varrho_{ikl} = 0 \qquad (A.11.29a)$$

but

$$\beta_{ikl} \neq 0 \qquad (A.11.29b)$$

which means that the torsion is nonzero:

$$A_{ikl} - A_{ilk} = 2\tau_{ikl} \neq 0 \qquad (A.11.30)$$

The curvature tensor, which we denote in the present instance by $G^i{}_{klm}$, is given by

$$G^i{}_{klm} = -A^i{}_{kl,m} + A^i{}_{km,l} - A^r{}_{kl}A^i{}_{rm} + A^r{}_{km}A^i{}_{rl}$$

$$= R^i{}_{klm} - \Omega^i{}_{kl;m} + \Omega^i{}_{km;l} - \Omega^r{}_{kl}\Omega^i{}_{rm} + \Omega^r{}_{km}\Omega^i{}_{rl} \quad \text{(A.11.31)}$$

This tensor does not in general vanish, and it exhibits only the following symmetries:

$$G_{iklm} = -G_{ikml} = -G_{kilm} \quad \text{(A.11.32)}$$

A comparison of displaced proper magnitudes is still possible:

$$(B^i B_i)_{;l}\frac{dx^l}{d\tau} = 0 \quad \text{(A.11.33)}$$

for equation (A.11.26) again obtains.

Under special circumstances, the curvature tensor $G^i{}_{klm}$ does vanish, which allows for comparisons between directions and not only between proper magnitudes. The displacement is then known as *Einsteinian parallel displacement over a distance*.

To understand what is here involved, let us introduce an orthonormal field of n-pods with contravariant components $h_D{}^i$ and covariant components h_{Dk} $(D = 1, 2, \ldots, n)$ and write the metric as

$$g_{ik} = h^D{}_i h_{Dk} \quad \text{(A.11.34)}$$

For this particular field, *Einstein's lemma*,

$$h^D{}_{i;l} \equiv h^D{}_{i,l} - h^D{}_k A^k{}_{il} = 0 \quad \text{(A.11.35)}$$

applies. Ricci's lemma is, of course, also valid. From Einstein's lemma, the Einstein coefficients, now denoted by $\Delta^i{}_{kl}$, of the Einstein displacement are then found to be

$$\Delta^i{}_{kl} = A^i_{kl} = h^{Di}h_{Dk,l} = h^{Di}{}_{,l}h_{Dk} \qquad (A.11.36)$$

or

$$\Delta_{ikl} = \Gamma_{ikl} + h^D{}_i h_{Dk;l} \qquad (A.11.37)$$

On comparison of this form with equation (A.21.11), the β_{ikl}, now called *Ricci's rotation coefficients* and designated by γ_{ikl}, are seen to be

$$\gamma_{ikl} = h^D{}_i h_{Dk;l} = -h^D{}_{i;l} h_{Dk} \qquad (A.11.38)$$

while the third-rank torsion tensor becomes

$$\tau_{ikl} = \tfrac{1}{2} h^D{}_i (h_{Dk,l} - h_{Dl,k}) \qquad (A.11.39)$$

Einstein's lemma guarantees the integrability of the displacements, the integrability condition being simply the vanishing of the curvature

$$G^i{}_{klm} = -\Delta^i{}_{kl,m} + \Delta^i{}_{km,l} - \Delta^r{}_{kl}\Delta^i{}_{rm} + \Delta^r{}_{km}\Delta^i{}_{rl} = 0 \qquad (A.11.40)$$

(iii) In the most general case one assumes, besides $\beta_{ikl} \neq 0$, that

$$\varrho_{ikl} \neq 0 \Leftrightarrow \Omega_{ikl} + \Omega_{kil} \neq 0 \qquad (A.11.41)$$

This assumption renders impossible any comparison not only

between directions but also between proper magnitudes, for now the square of B^i is found to change along the line $x^l = x^l(\tau)$ in accordance with

$$B^i B^k g_{ik;l} = -2\varrho_{ikl} B^i B^k \frac{dx^l}{d\tau} \qquad \text{(A.11.42)}$$

It follows from this that the absolute values themselves of vectorial properties of systems are found to be different if they are analyzed along different lines of approach from an initial position. This means that the proper magnitudes assigned to a physical system, embedded in space–time, depend on its past history. Thus, suppose that two particles possess the same proper magnitudes when at time $t = 0$ they occupy the same point P_0 of the subspace V_3, but that they are from then on transported, according to Levi–Civita's definition of a parallel displacement, along two different lines. Then, if their paths were to intersect in the point P' at some later time t', the proper magnitudes of the two systems would no longer agree. These magnitudes possess biogrammatic features, which one can interpret by saying that the part of Ω_{ikl} that is symmetric in the first two indices, viz., ϱ_{ikl}, incorporates the temporal (historical) behavior of the physical processes. This was discovered by Einstein as early as 1918 in a discussion with Hermann Weyl on conformal geometry.

The only symmetry of the Riemann tensor that survives in the general case (iii) is

$$G_{iklm} = -G_{ikml} \qquad \text{(A.11.43)}$$

No longer valid is the antisymmetry of G_{iklm} with respect to its first two indices, which guaranteed integrability in combination with Ricci's lemma.

Appendix B

General Relativity Principle
and General Lorentz Covariance

B.1. Lorentz-Covariant Derivatives

It is essential for the understanding of the general relativity principle to distinguish between coordinate systems $\{x^i\}$ and reference systems or frames Σ: *Coordinate systems* are purely mathematical constructs for the presentation of mathematical relations; and the independence of the measured values of physical quantities from the choice of the coordinate system $\{x^i\}$, although logically necessary, has no physical consequences. *Reference systems*, by contrast, are physically real; they correspond to a set of measurement instruments designed for the determination of the physical quantities. In the simplest case the measurements in question can be realized with the aid of three standard rods and a standard clock. Thus, one associates three measuring rods and a clock with each event in space–time (Treder, 1966).

From a mathematical point of view, a reference system Σ is represented by a field of the *tetrads* $h^A{}_i$, which one may assume to be orthonormalized:

$$g_{ik} = h^A{}_i h^B{}_k \eta_{AB}$$

$$\eta_{AB} = h^i{}_A h^k{}_B g_{ik} \tag{B.1.1}$$

Here g_{ik} is the space–time *metric tensor* and

$$(\eta_{AB}) = \mathrm{diag}(1, -1, -1, -1) \tag{B.1.2}$$

the *Minkowski tensor*. The lower-case Latin indices identify tensor components and the capital indices number the four space–time vectors; both have the range 0, 1, 2, 3. The $h^A{}_i$ are functions of the space–time coordinates.

Equation (B.1.1) expresses the condition for the compatibility of a reference system $h^A{}_i$ with a space–time V_4 of given metric g_{ik}. A space–time transformation

$$x'^l = x'^l(x^k)$$

$$h'^A{}_l = \frac{\partial x^k}{\partial x'^l} h^A{}_k \tag{B.1.3}$$

of the vector tetrad belonging to the Einstein group gives rise to the *isometric metric*

$$g'_{mn} = \frac{\partial x^k}{\partial x'^m} \frac{\partial x^l}{\partial x'^n} h^A{}_k h_{Al}$$

$$= \frac{\partial x^k}{\partial x'^m} \frac{\partial x^l}{\partial x'^n} g_{kl} \tag{B.1.4}$$

Equation (B.1.1) associates a universal *Minkowskian tangential space* M_4 with the space–time V_4. This M_4 is the *dual manifold*

V_4^* of V_4. The transformation matrix $h^A{}_k$ that connects V_4 with V_4^* is in general anholonomic with the *Einsteinian anholonomic object* (Einstein, 1928; Schouten, 1953):

$$\Delta^i_{\underset{\vee}{k}l} = \tfrac{1}{2} h^i{}_A (h^A{}_{k,l} - h^A{}_{l,k}) \tag{B.1.5}$$

It should be noted that, together with $h^A{}_i$, also the Lorentz-rotated tetrad field

$$\bar{h}^B{}_i = \omega^B{}_A h^A{}_i \tag{B.1.6}$$

with

$$\omega_A{}^C \omega_{BC} = \eta_{AB} \tag{B.1.7}$$

will furnish a reference system that is compatible with the given metric g_{ik}. The *principle of general relativity* now asserts that all reference systems Σ that are compatible with the given metric structure g_{ik} of the space–time V_4 are equivalent.

The geometrical objects of space–time V_4 fulfill these presuppositions precisely if and only if they depend only on the Lorentz-invariant combination (B.1.1) of the tetrads and their derivatives (see below). All Lorentz invariants are tensors with respect to transformations of the Einstein group.

The measured values ϕ^I of physical quantities are invariant under the choice of coordinate system—that is, they must be pure point functions of the coordinates and consequently space–time scalars:

$$\phi'^I = \phi^I(x^i(x'^k)) = \phi^I(x^i) \tag{B.1.8}$$

According to the relativity principle, it is, however, further required that the relations among the measured values ϕ^I be independent of the choice of the reference system as well. In particular, $\bar{\phi}^I = 0$ should be valid whenever $\phi^I = 0$. It follows that the matrix ϕ^I must behave covariantly in relation to the Lorentz transformation (B.1.7), that is, the matrix ϕ^I of the measured value must be a Lorentz tensor of the rank n:

$$\bar{\phi}_{B_1\ldots}^{A_1\ldots} = \omega^{A_1}{}_{C_1} \cdots \omega_{B_1}{}^{D_1} \cdots \phi_{D_1\ldots}^{C_1\ldots} \tag{B.1.9}$$

From the last two equations in conjunction with equation (B.1.1), it follows that one can uniquely associate, with all measured values of physical quantities, space–time tensors of the same rank,

$$\phi_{k_1\ldots}^{i_1\ldots} = h^{i_1}{}_{A_1} \cdots h^{B_1}{}_{k_1} \cdots \phi_{B_1\ldots}^{A_1\ldots} \tag{B.1.10}$$

which, in their turn, are Lorentz-invariant. Lorentz tensors $\phi_{B_1\ldots}^{A_1\ldots}$ and space–time tensors $\phi_{k_1\ldots}^{i_1\ldots}$ are dual quantities.

The measured values ϕ^A of a quantity ϕ^i in the reference system $h^A{}_i$ are constant in the space–time V_4 if the change of ϕ^i satisfies the equation

$$\phi^i{}_{,l} + h^i{}_A h^{-A}{}_{k,l}\phi^k = \phi^i{}_{,l} + \Delta^i{}_{kl}\phi^k = 0 \tag{B.1.11}$$

Here the symbols

$$\Delta^i{}_{kl} = h^i{}_A h^A{}_{k,l} \tag{B.1.12}$$

denote the *affine coefficients of Einstein's* (1928) *distant parallel (integrable) transfer*, which obey the identity

$$h^A{}_{i,l} - \Delta^k{}_{il}h^A{}_k = 0 \tag{B.1.13}$$

The ordinary differential $d\phi^{i\cdots}_{k_1\cdots}$ of a space–time tensor is not a tensor, since, for example,

$$d\phi'^i = \phi'^i{}_{,l}\, dx^l$$

$$= \left(\frac{\partial x'^i}{\partial x^r}\,\phi^r{}_{,l} + \frac{\partial^2 x'^i}{\partial x^r\,\partial x^l}\,\phi^r\right) dx^l \qquad (B.1.14)$$

A tensor can, however, be constructed through addition of a compensating term

$$\Gamma^i_{kl}\phi^k\, dx^l \qquad (B.1.15)$$

where the *affine connection* or *affinity* Γ^i_{kl} obeys the transformation law

$$\Gamma'^i_{kl} = \frac{\partial x'^i}{\partial x^m}\,\frac{\partial x^r}{\partial x'^k}\,\frac{\partial x^p}{\partial x'^l}\,\Gamma^m_{rp} + \frac{\partial x'^i}{\partial x^m}\,\frac{\partial^2 x^m}{\partial x'^k\,\partial x'^l} \qquad (B.1.16)$$

The only affinity that can be constructed from g_{ik} and its derivatives is the *Christoffel three-index symbol*

$$\left\{{i \atop kl}\right\} = \frac{1}{2}\,g^{ir}(-g_{kl,r} + g_{lr,k} + g_{rk,l}) \qquad (B.1.17)$$

It is evidently a Lorentz invariant.

Likewise, the (coordinate-invariant) ordinary differential

$$d\phi^A = \phi^A{}_{,l}\, dx^l$$

$$= \phi^A{}_{,l}h^l{}_B\, dx^B$$

$$= \phi^A{}_{,B}\, dx^B \qquad (B.1.18)$$

of a Lorentz vector ϕ^A is not a Lorentz tensor, for

$$d\bar{\phi}^A = (\omega^A{}_B\phi^B)_{,l}\,dx^l$$

$$= (\omega^A{}_B\phi^B{}_{,l} + \omega^A{}_{B,l}\phi^B)\,dx^l \qquad (B.1.19)$$

One must therefore form a Lorentz-covariant derivative by introducing a compensating *Lorentz affinity*

$$L^A_{Bl}\phi^B\,dx^l = L^A_{BC}\phi^B\,dx^C \qquad (B.1.20)$$

with the transformation law

$$\bar{L}^A_{Bl} = \omega^A{}_C\omega_B{}^D L^C_{Dl} + \omega^A{}_D\omega_B{}^D{}_{,l} \qquad (B.1.21)$$

The only Lorentz affinity that can be constructed from the tetrads and their derivatives is the quantity

$$L^A_{Bl} = -\gamma^A_{Bl} = h^A{}_i h^i{}_{B;l} = -h^A{}_{i;l}h^i{}_B \qquad (B.1.22)$$

which indeed has the transformation law

$$\bar{\gamma}^A_{Bl} - \omega^A{}_C\omega_B{}^D\gamma^C_{Dl} = -\omega^A{}_D\omega_B{}^D{}_{,l} = \omega^A{}_{D,l}\omega_B{}^D \qquad (B.1.23)$$

[If in a flat V_4 we go out from a special-relativistic inertial system, for which in Cartesian coordinates $h^A{}_i = \delta^A{}_i$, and carry out general Lorentz transformations $h^A{}_i = \omega^A{}_B\delta^B{}_i$, then we find

$$\gamma^A_{Bl} = -\omega^A{}_C\omega_B{}^C{}_{,l}$$

186

(see below).] Equation (B.1.22) defines the *Ricci rotation coefficients* for the tetrad field $h^A{}_i$. The space–time components of the Ricci rotation,

$$\gamma^i_{kl} = h^i{}_A h^B{}_k \gamma^A_{Bl} = h^i{}_A h^A{}_{k;l} = -h^i{}_{A;l} h^A{}_k \qquad (B.1.24)$$

with

$$\phi^i{}_{;k} = \phi^i{}_{,k} + \phi^l \begin{Bmatrix} i \\ kl \end{Bmatrix} \qquad (B.1.25)$$

are by themselves not legitimate general-relativistic quantities, because they indeed form a space–time tensor, but not a Lorentz scalar. In this respect they may be contrasted with the Christoffel affinity $\begin{Bmatrix} i \\ kl \end{Bmatrix}$, which indeed is a Lorentz scalar but not a space–time tensor.

With the help of the Lorentz affinity (B.1.22) and using the Leibniz rule, we define a *Lorentz-covariant derivative* of, for example, a mixed Lorentz tensor $\phi^A{}_B$ as

$$\phi^A{}_{B\|C} = \phi^A{}_{B\|l} h^l{}_C$$

$$= \phi^A{}_{B,l} - \gamma^A_{DC} \phi^D{}_B + \gamma^D_{BC} \phi^A{}_D \qquad (B.1.26)$$

For a mixed space–time and Lorentz tensor, e.g., $\phi^A{}_i$, we define, with the aid of the Ricci rotation coefficients and the Christoffel affinity, the *generally-covariant* (Lorentz-covariant and coordinate-covariant) *derivative* as

$$\phi^A{}_{i\|l} = \phi^A{}_{i,l} - \phi^B{}_i \gamma^A_{Bl} - \phi^A{}_r \begin{Bmatrix} r \\ il \end{Bmatrix} \qquad (B.1.27)$$

Appendix B

This derivative reduces to the coordinate-covariant derivative

$$\phi^i{}_{\| l} = \phi^i{}_{,l} + \left\{{i \atop rl}\right\}\phi^r$$

$$= \phi^i{}_{;l} \tag{B.1.28}$$

for pure space–time quantities and to the Lorentz-covariant derivative

$$\phi^A{}_{\| l} = \phi^A{}_{,l} - \gamma^A{}_{Bl}\phi^B$$

$$= \phi^A{}_{\| l} \tag{B.1.29}$$

for pure Lorentz tensors.

Apart from the known identities

$$\delta^i{}_{k;l} = \delta^i{}_{k,l} = 0 \tag{B.1.30a}$$

$$\delta^A{}_{B\| l} = \delta^A{}_{B,l} = 0 \tag{B.1.30b}$$

$$g_{ik\| l} = g_{ik;l} = 0 \tag{B.1.30c}$$

$$\eta_{AB\| l} = \eta_{AB\| l} = \gamma_{ABl} + \gamma_{BAl} = 0 \tag{B.1.30d}$$

we moreover find that the tetrads $h^A{}_i$ are generally-covariant constant:

$$h^A{}_{i\| l} = h^A{}_{i,l} - h^B{}_i\gamma^A{}_{Bl} - h^A{}_r\left\{{r \atop il}\right\}$$

$$= h^A{}_{i,l} - h^A{}_{i;l} - h^A{}_r\left\{{r \atop il}\right\} = 0 \tag{B.1.31}$$

which is Weyl's lemma (Weyl, 1929). In fact,

$$\gamma^i_{kl} + \begin{Bmatrix} i \\ kl \end{Bmatrix} = \Delta^i_{kl}$$

$$= h^i{}_A h^A{}_{k,l} \tag{B.1.32}$$

expresses again the Einstein affinity with distant parallelism (Einstein, 1928).

[Weyl's lemma, $h^A{}_{i\|l} = 0$, must be valid if one requires the equivalence of the Lorentz and Einstein representations. As a consequence, the connection between Γ^i_{kl} and $L^A{}_{Bl}$ is given by the anholonomic transformation (B.1.1): $\Gamma^i_{kl} = L^i_{kl} + h^i{}_A h^A{}_{k,l}.$]

By virtue of equations (B.1.1) and (B.1.31), a one-to-one correspondence (duality) exists between the Lorentz-covariant and the coordinate-covariant representations of tensor quantities. So, for example, *Eisenhart's identity* (Eisenhart, 1927),

$$\phi^A{}_{\|C} = (h^A{}_i \phi^i)_{\|l} h^l{}_C$$

$$= \phi^i{}_{;l} h^A{}_i h^l{}_C \tag{B.1.33a}$$

holds as well as

$$\phi^A{}_{\|CD} = \phi^i{}_{;kl} h^A{}_i h^k{}_C h^l{}_D \tag{B.1.33b}$$

etc. One can therefore replace the Lorentz-covariant derivatives of Lorentz tensors by the coordinate-covariant derivatives of space–time tensors, and conversely. The two types of covariant derivatives are dual to one another.

In particular, the Lorentz-covariant curvature tensor is identical with the curvature tensor of Riemann.

One finds that

$$\phi^A{}_{\|BC} - \phi^A{}_{\|CB} = (\phi^i{}_{;kl} - \phi^i{}_{;lk})h^A{}_i h^k{}_B h^l{}_C$$

$$= -R^i{}_{rkl}\phi^r h^A{}_i h^k{}_B h^l{}_C$$

$$= -\phi^D R^A{}_{DBC} \qquad (B.1.34)$$

The inertial law of special relativity,

$$u^l{}_{,k}u^k = 0 \qquad (B.1.35)$$

is capable of being expressed in, for example, Lorentz-covariant manner by four scalar equations:

$$u^A{}_{\|C}u^C = (u^A{}_{,i} - \gamma^A_{Bi}u^B)h^i{}_C u^C = 0 \qquad (B.1.36)$$

and leads consequently to the equation of a geodesic in a Riemannian V_4:

$$u^i{}_{;k}u^k h^A{}_i = u^A{}_{\|C}u^C = 0 \qquad (B.1.37)$$

Decisive for the derivation of the geodesic equation is the demand of covariance under arbitrary position-dependent Lorentz rotations. If one required only covariance relative to rigid Lorentz rotations $\omega^A{}_{B,l} = 0$, then there would be no need to augment the ordinary differential

$$d\phi^A = \phi^A{}_{,l}\,dx^l$$

by a Lorentz affinity:

$$\phi^A{}_{,l}\,dx^l = (\phi^i{}_{,l}h^A{}_i + h^A{}_{i,l}\phi^i)\,dx^l \qquad (B.1.38)$$

Equation (B.1.38) leads to its dual coordinate-covariant derivative

$$h^k_{,1} \, d\phi^{,1} = (\phi^k_{,l} + h^k_{,1} h^{,1}_{i,l} \phi^i) \, dx^l \qquad \text{(B.1.39)}$$

In these differentials the coordinate affinity is equal to the integrable Einstein affinity Δ^i_{kl}. The tensor transport in V_4 then is integrable; there exists no affine curvature and therefore no gravitational action. The principle of general relativity, by requiring the Lorentz covariance of the relations among the measured values of physical quantities, consequently implies the Einstein principle of the space–time covariance of physical equations in their pure space–time formulation and renders possible the geometrization of the gravitational field.

On the assumption that the structure of space–time is determined solely by the g_{ik} and that all measurable physical quantities are characterized completely by Lorentz tensors, it is thus possible to uniquely formulate physical equations in a pure space–time fashion, that is, in terms of pure space–time tensors, independently of any particular frame of reference. The general relativity principle then reads as follows (Treder, 1966):

The fundamental physical equations can be formulated in space–time V_4 without mention of any frame of reference (*Einstein's general relativity principle*).

In order that the relations among physical measurement values, which follow from these fundamental equations in space–time, be independent of any choice of coordinate system (and thus interpretable), it is necessary for the space–time equations to be coordinate covariant (*Einstein's covariance principle*). The Einstein form of the general relativity principle is evidently the dual version of the Lorentz covariance requirement (Treder, 1969).

B.2. Spinor Calculus

For tensor fields there exist two equivalent representations (dual to one another) which satisfy the general relativity principle (Treder, 1969):

1. The Lorentz-covariant representation, which employs coordinate-invariant quantities $\phi_{B\ldots}^{A\cdots}$.
2. The coordinate-covariant representation, which employs Lorentz-invariant quantities $\phi_{k\ldots}^{i\cdots}$.

The equivalence of these representations is a consequence of the uniqueness of the correspondences

$$\phi_{D\ldots}^{A\cdots} = h^A{}_i \cdots h^k{}_D \cdots \phi_{k\ldots}^{i\cdots} \tag{B.2.1a}$$

and

$$\phi_{k\ldots}^{i\cdots} = h^i{}_A \cdots h^D{}_k \cdots \phi_{D\ldots}^{A\cdots} \tag{B.2.1b}$$

Such a unique correspondence is no longer possible when spinor quantities are used. The introduction of spinor quantities means nothing more for tensor fields than the replacement of the one-valued tensor representation of the Lorentz group by the two-valued unimodular representation. In this process, one associates metric spin vectors with the tetrads of the reference system through

$$\sigma^{l\mu\dot{\nu}} = h^l{}_A \sigma^{A\mu\dot{\nu}} \tag{B.2.2}$$

where the $\sigma^{A\mu\dot{\nu}}$ stand for the constant Pauli spin matrices, and the Greek, spinor indices can assume the values μ, $\nu = 1$, 2 ($\varphi^{\dot{\mu}}$ denotes the complex conjugate to φ^μ).

The orthogonality condition (B.1.1) now reads

$$\gamma_{\alpha\beta}\gamma_{\dot\mu\dot\nu} = \sigma^k{}_{\alpha\dot\mu}\sigma^l{}_{\beta\dot\nu}g_{kl}$$

$$= \sigma^A{}_{\alpha\dot\mu}\sigma^B{}_{\beta\dot\nu}\eta_{AB} \tag{B.2.3a}$$

and

$$g_{kl} = \sigma_k{}^{\alpha\dot\mu}\sigma_l{}^{\beta\dot\nu}\gamma_{\alpha\beta}\gamma_{\dot\mu\dot\nu}$$

$$= h^A{}_k h^B{}_l \sigma_A{}^{\alpha\dot\mu}\sigma_B{}^{\beta\dot\nu}\gamma_{\alpha\beta}\gamma_{\dot\mu\dot\nu} \tag{B.2.3b}$$

Here the quantities

$$\gamma_{\alpha\beta} = -\gamma_{\beta\alpha} = \gamma_{\dot\alpha\dot\beta} = -\gamma_{\dot\beta\dot\alpha} \tag{B.2.4a}$$

with

$$|\gamma_{\alpha\beta}| = 1, \qquad \gamma_{\alpha\beta}\gamma^{\varepsilon\beta} = \delta_\alpha{}^\varepsilon$$

are the metric tensors of spin space and according to (B.2.3) the constants

$$\gamma_{1i} = -\gamma_{21} = -\gamma^{12} = \gamma^{21} = 1 \tag{B.2.4b}$$

The representation (B.2.3) is invariant under the unimodular transformation in spin space S_2 or S_2^*, that is, under transformations

$$\bar\sigma_i{}^{\alpha\dot\mu} = \alpha^{\dot\mu}{}_{\dot\nu}\alpha^\alpha{}_\beta\sigma_l{}^{\beta\dot\nu} \tag{B.2.5}$$

of the metric spin vectors, for which the transformation matrices $\alpha^\mu{}_\nu = \alpha^\mu{}_\nu(x^l)$ satisfy the condition

$$|\alpha^{\dot\mu}{}_{\dot\nu}| = |\alpha^\mu{}_\nu| = 1 \tag{B.2.6}$$

Appendix B

If the measured quantities are considered in relation to the spin spaces S_2 and S_2^*, then the general relativity principle requires —corresponding to the unimodular invariance of (B.2.3)—that all these quantities transform like spinors:

$$\bar{\psi}^\nu = \alpha^\nu{}_\mu \psi^\mu$$

$$\bar{\psi}^{\dot{\nu}} = \alpha^{\dot{\nu}}{}_{\dot{\mu}} \psi^{\dot{\mu}}$$

(B.2.7)

In view of equation (B.2.2), this condition is equivalent, for Hermitian spinors of rank $2n$, e.g., for

$$\psi_{\mu\dot{\nu}}, \qquad \psi_{\mu\dot{\nu}\alpha\beta}$$

etc., with the demand that the measured quantities be Lorentz tensors:

$$\phi_{\mu\dot{\nu}} = \sigma_{A\mu\dot{\nu}} \phi^A$$

$$= \sigma_{i\mu\dot{\nu}} h^i{}_A \phi^A$$

$$= \sigma_{i\mu\dot{\nu}} \phi^i$$

If ψ_ν is a spinor, then the ordinary derivative

$$d\psi_\nu = \psi_{\nu,l}\, dx^l$$

(B.2.8)

is not a spinor, since one gets

$$\bar{\psi}_{\nu,l} = (\alpha_\nu{}^\mu \psi_\mu)_{,l}$$

$$= \alpha_\nu{}^\mu \psi_{\mu,l} + \alpha_\nu{}^\mu{}_{,l} \psi_\mu$$

(B.2.9)

General Relativity Principle and General Lorentz Covariance

By introducing a compensating *spinor affinity* $\Lambda^{\alpha}_{\beta l}$, with the transformation law

$$\bar{\Lambda}^{\alpha}_{\beta l} = \alpha_{\beta}{}^{\mu}\alpha^{\alpha}{}_{\nu}\Lambda^{\nu}_{\mu l} + \alpha^{\alpha}{}_{\lambda}\alpha_{\beta}{}^{\lambda}{}_{,l} \tag{B.2.10}$$

one can, however, construct the *Lorentz-covariant spinor derivative*

$$\psi_{\nu \| l}\, dx^{l} = (\psi_{\nu,l} - \Lambda^{\mu}_{\nu l}\psi_{\mu})\, dx^{l} \tag{B.2.11}$$

As for the definition of the spinor affinity (B.2.10), one is guided by the consideration that it should at the same time be a space–time tensor to guarantee that the spinor derivative (B.2.11) will exhibit coordinate-covariant behavior. The only affinity of this nature, which can be constructed from the spin vectors and their derivatives alone, is the *spinor affinity of Infeld and van der Waerden* (1933):

$$\Lambda^{\alpha}_{\beta l} = \tfrac{1}{2}\sigma^{i\alpha\dot{\nu}}\sigma_{i\beta\dot{\nu};l}$$

$$= \tfrac{1}{2}\sigma^{i\alpha}{}_{\dot{\nu};l}\sigma_{i\beta}{}^{\dot{\nu}} \tag{B.2.12}$$

It is again possible to define *generally-covariant derivatives* for mixed-tensor and mixed-spinor quantities; so, for example,

$$\phi^{k}{}_{\nu \| l} = \phi^{k}{}_{\nu,l} + \begin{Bmatrix} k \\ rl \end{Bmatrix}\phi^{r}_{\nu} - \Lambda^{\alpha}{}_{\nu l}\phi^{k}{}_{\alpha} \tag{B.2.13}$$

On the basis of (B.2.13), one obtains the known generally-covariant constancy of the metric spin vectors

$$\sigma^{k\mu\dot{\nu}}{}_{\| l} = \sigma^{k\mu\dot{\nu}}{}_{,l} + \sigma^{r\mu\dot{\nu}}\begin{Bmatrix} k \\ rl \end{Bmatrix} + \sigma^{k\alpha\dot{\nu}}\Lambda^{\mu}{}_{\alpha l} + \sigma^{k\mu\dot{\alpha}}\Lambda^{\dot{\nu}}{}_{\dot{\alpha} l} = 0 \tag{B.2.14}$$

Indeed, one has

$$\sigma^{k\mu\dot{\nu}}{}_{\|l} = \sigma^{k\mu\dot{\nu}}{}_{,l} + \sigma^{r\mu\dot{\nu}} \Lambda^k_{rl} = 0 \qquad (B.2.15)$$

where Λ^k_{rl} is the Einstein affinity (B.1.32),

$$\Lambda^k_{rl} = h^k{}_A h^A{}_{r,l}$$

$$= \sigma^{k\mu\dot{\nu}} \sigma_{r\mu\dot{\nu},l} \qquad (B.2.16)$$

With reference to (B.2.4) and (B.2.20), it follows that

$$\gamma_{\alpha\beta\|l} = -\gamma_{\beta\alpha\|l}$$

$$= \gamma_{\alpha\lambda} \Lambda^\lambda_{\beta l} + \gamma_{\lambda\beta} \Lambda^\lambda_{\alpha\beta}$$

$$= \Lambda_{\alpha\beta l} - \Lambda_{\beta\alpha l} \qquad (B.2.17)$$

There exists a duality between the coordinate-covariant equations for tensor fields and the Lorentz-covariant equations for the corresponding spinor fields of even rank:

$$\phi_{\mu\dot{\nu}\|l} = (\sigma_{i\mu\dot{\nu}} \phi^i)_{\|l}$$

$$= \sigma_{i\mu\dot{\nu}} \phi^i{}_{;l} \qquad (B.2.18)$$

From equations (B.1.27) and (B.2.2), one furthermore finds, because of

$$\phi_{\mu\dot{\nu}} = \phi^k \sigma_{k\mu\dot{\nu}}$$

$$= \phi^A \sigma_{A\mu\dot{\nu}} \qquad (B.2.19)$$

that

$$\phi_{\mu\dot\nu\|l} = \phi^A{}_{\|l}\sigma_{A\mu\dot\nu} + \phi^A\sigma_{A\mu\dot\nu\|l} \qquad (B.2.20)$$

However, equations (B.1.30) and (B.2.14) give rise to (Einstein and Mayer, 1932):

$$\sigma^{A\mu\dot\nu}{}_{\|l} = (\sigma^{i\mu\dot\nu}h^A{}_i)_{\|l}$$

$$= \sigma^{i\mu\dot\nu}{}_{\|l}h^A{}_i + \sigma^{i\mu\dot\nu}h^A{}_{i\|l} = 0 \qquad (B.1.21)$$

It consequently follows from equations (B.2.12) and (B.2.10) that

$$\phi_{\mu\dot\nu\|l}\sigma^{A\mu\dot\nu} = \phi^A{}_{\|l}$$

$$= h^A{}_i\phi^i{}_{;l} \qquad (B.2.22)$$

Spinor derivative and Lorentz-covariant derivative are thus equivalent when applied to the even-rank spinors that are dual to tensor fields. By contrast, it is not possible to transform Lorentz-covariant equations for spinor fields of odd rank into coordinate-covariant equations without spinor indices.

As on epistemological grounds should be the case, the equations for arbitrary spinor fields are therefore coordinate invariant and, according to the relativity principle, amenable to Lorentz-covariant formulation. For odd-rank spinors, however, no Lorentz-invariant and coordinate-covariant representation exists. Correspondingly, the general version of the principle of general relativity is contained in the statement that all equations of physics can be formulated in a Lorentz-covariant and coordinate-invariant manner. For tensor quantities, there then still exists the dual version that quantities occurring in physics should permit Lorentz-invariant and coordinate-covariant formulations.

By means of the coordinate-invariant and Lorentz-covariant formulation of physical equations, it becomes possible to fulfill the general relativity principle for spinor as well as tensor fields. A violation of the general principle, in the sense of the general Lorentz covariance of the relations among measurable physical quantities, can only occur if it should happen that the geometry of space–time V_4 is determined not only by the Lorentz-invariant metric g_{ik}, but also by combinations of the tetrad fields $h^{\cdot 1}{}_k$ that are not Lorentz-invariant. This means that the laws determining the structure of space–time must naturally always still be coordinate-covariant. If these equations fix only the g_{ik}, then the structure of the space is Lorentz-invariant. But if these equations determine also other combinations that are not Lorentz-invariant, then the effect is to break the general Lorentz covariance of the geometrical structure of V_4. In particular, the general Lorentz covariance of the structure of V_4 is completely destroyed when, as in the tetrad theory, all 16 components of the tetrad field $h^{\cdot 1}{}_k$ are determined by the structure equations of space–time V_4 (Treder, 1967, 1971).

References

References for the Introduction

EINSTEIN, A. (1907). *Jahrb. Radioakt.* **4**, 411.

EINSTEIN, A. (1917). *Sitzungsber. Preuss. Akad. Wiss.*, 142.

HERTZ, H. (1894). *Die Prinzipien der Mechanik*, J. A. Barth, Leipzig.

MACH, E. (1883). *Die Mechanik in ihrer Entwicklung*, F. A. Brockhaus, Leipzig.

TREDER, H.-J. (1972). *Die Relativität der Trägheit*, Akademie-Verlag, Berlin.

References for Chapter 1

ABRAHAM, M. (1920). *Theorie der Elektrizität*, Vol. 2, 4th ed., B. G. Teubner, Leipzig.

ANDING, E. (1905). In *Enzyklopädie der Mathematischen Wissenschaften*, Vol. IV, Part 2, page 3.

References

AOIKI, S. (1967). *Publ. Astron. Soc. Jpn.* **19**, 585.

BIRKHOFF, G. D. (1923). *Relativity and Modern Physics*, University Press, Cambridge.

BOPP, F. (1940). *Ann. Phys. (Leipzig)* **38**, 345.

EINSTEIN, A. (1913). *Phys. Z.* **14**, 1249.

EINSTEIN, A. (1914). *Sitzungsber. Preuss. Akad. Wiss.*, 1030.

EINSTEIN, A. (1915). *Sitzungsber. Preuss. Akad. Wiss.*, 778.

EINSTEIN, A. (1916a). *Sitzungsber. Preuss. Akad. Wiss.*, 1111.

EINSTEIN, A. (1916b). *Die Grundlagen der Allgemeinen Relativitäts-theorie*, J. A. Barth, Leipzig.

EINSTEIN, A. (1918). *Sitzungsber. Preuss. Akad. Wiss.*, 154.

EINSTEIN, A. (1921). *Geometrie und Erfahrung*, J. Springer, Berlin.

EINSTEIN, A., AND FOKKER, A. D. (1914). *Ann. Phys. (Leipzig)* **44**, 321.

EINSTEIN, A., AND GROSSMANN, M. (1913). *Entwurf einer Verall-gemeinerten Relativitätstheorie und eine Theorie der Gravitation*, B. G. Teubner, Berlin–Leipzig.

EINSTEIN, A., AND MAYER, W. (1932). *Sitzungsber. Preuss. Akad. Wiss.*, 522.

EINSTEIN, A., AND PAULI, W. (1943). *Ann. Math.* **44**, 131.

FOCK, V. A. (1960). *Theorie von Raum, Zeit und Gravitation*, Akademie-Verlag, Berlin.

FOKKER, A. D. (1920). *Proc. Acad. Sci. Amsterdam* **23**, 729.

FOKKER, A. D. (1965). *Time and Space, Weight and Inertia*, Pergamon Press, Elmsford (New York).

HERTZ, H. (1894). *Die Prinzipien der Mechanik*, Leipzig.

HILBERT, D. (1915). *Die Grundlagen der Physik*, Vols. 1 and 2; *Göttinger Nachr.*, 395 (1915).

JÁNOSSY, L., AND TREDER, H.-J. (1972). *Acta Phys. Hung.* **31**, 367.

KOHLER, M. (1952). *Z. Phys.* **131**, 571.

KOHLER, M. (1953). *Z. Phys.* **134**, 286.

References

VON LAUE, M. (1917). *Jahrb. Radioakt. Elektron.* **14**, 263.

VON LAUE, M. (1956). *Die Relativitätstheorie*, Vol. 2, 4th ed., F. Vieweg, Braunschweig.

MIE, G. (1923), *Ann. Phys. (Leipzig)* **70**, 489.

MINKOWSKI, H. (1908). *Göttinger Nachr.* **53**.

MINKOWSKI, H. (1909). *Phys. Z.* **10**, 104.

NORDSTRÖM, G. (1913). *Ann. Phys. (Leipzig)* **42**, 533.

PLANCK, M. (1907). *Sitzungsber. Preuss. Akad. Wiss.*, 542.

PODOLSKY, B. (1941). *Phys. Rev.* **62**, 68.

POINCARÉ, H. (1906). *R. C. Circ. Mat. Palermo* **21**, 129.

POINCARÉ, H. (1912). *Wissenschaft und Hypothese*, 3rd ed., B. G. Teubner, Leipzig.

POINCARÉ, H. (1914). *Wissenschaft und Methode*, B. G. Teubner, Leipzig.

ROSEN, N. (1940). *Phys. Rev.* **57**, 147, 150.

VON SEELIGER, H. (1906). *Sitzungsber. Bayr. Akad. Wiss.* **36**, 85.

DE SITTER, W. (1916). *Mon. Not. R. Astron. Soc.* **77**, 155.

TOLMAN, R. C. (1934). *Relativity, Thermodynamics and Cosmology*, University Press, Oxford.

TREDER, H.-J. (1967). *Ann. Phys. (Leipzig)* **20**, 194.

TREDER, H.-J. (1970). *Found. Phys.* **1**, 75.

TREDER, H.-J. (1971a). *Monatsber. Deutsch. Akad. Wiss.* **13**, 310.

TREDER, H.-J. (ed.) (1971b). *Gravitationstheorie und Äquivalenzprinzip*, Akademie-Verlag, Berlin.

TREDER, H.-J. (1973). *Ann. Phys. (Leipzig)* **29**, 233.

TREDER, H.-J. (1974). *Exp. Tech. Phys.* **22**, 1.

TREDER, H.-J. (1975). *Ann. Phys. (Leipzig)* **32**, 383.

WEYL, H. (1923). *Raum—Zeit—Materie*, 5th ed., Berlin.

WEYL, H. (1924). *Naturwissenschaften* **12**, 197.

WEYL, H. (1929a). *Proc. Nat. Acad. Sci. (U.S.A.)* **15**, 323.

WEYL, H. (1929b). *Z. Phys.* **56**, 330.

References for Chapter 2

For convenience, the references in this chapter have been grouped by section.

2.1. Significance of the Post-Newtonian Kepler Problem

CHAZY, A. (1928, 1930). *La Théorie de la Relativité et la Mécanique Céleste*, Vols. 1 and 2, Gauthier-Villars, Paris.

EINSTEIN, A. (1915). *Sitzungsber. Preuss. Akad. Wiss.*, 831.

EINSTEIN, A. (1916). *Grundlage der Allgemeinen Relativitätstheorie*, J. A. Barth, Leipzig.

EINSTEIN, A. (1948). German original text from "Autobiographical Notes," in P. A. Schilpp (ed.), *Albert Einstein: Philosopher-Scientist*, Open Court Publishing Company, La Salle–Chicago (1949), p. 76.

EINSTEIN, A., AND INFELD, L. (1940). *Ann. Math.* **41**, 455.

EINSTEIN, A., AND INFELD, L. (1949). *Can. J. Math.* **1**, 209.

EINSTEIN, A., INFELD, L., AND HOFFMANN, B. (1938). *Ann. Math.* **39**, 65.

FOCK, V. A. (1955). *Teorija Prostranstwa, Vremeni i Tjagotenija*, Fizmatizdat, Moscow (in Russian) [German translation: (1960). *Theorie von Raum, Zeit und Gravitation*, Akademie-Verlag, Berlin].

GERBER, P. (1898). *Z. Math. Phys.* **43**, 93.

GERBER, P. (1917). *Ann. Phys. (Leipzig)* **52**, 415.

HALL, A. (1895). *Astron. J.* **24**, 49.

HARZER, P. (1891). *Astron. Nachr.* **127**, 82.

HARZER, P. (1896). *Astron. Nachr.* **141**, 39.

ISENKRAHE, C. (1879). *Das Rätsel von der Schwerkraft*, F. Vieweg, Braunschweig.

References

DE LAPLACE, P. S. (1798–1825). *Traité de Mécanique Céleste,* Bachelier, Paris; also in *Oeuvres,* Vols. I–V, Gauthier-Villars, Paris (1878–1883).

LEVERRIER, U. (1859). *Ann. Obs. Imperial* **5**, 78.

LEVI-CIVITA, T. (1947). *The Absolute Differential Calculus,* Blackie and Son, London–Glasgow. Later edition 1950.

MACH, E. (1912). *Die Mechanik in ihrer Entwicklung,* F. A. Brockhaus, Leipzig.

NEUMANN, C. (1868). *Über die Prinzipien der Elektrodynamik,* Universitäts-Buchhandlung, Tübingen.

NEUMANN, C. (1896). *Allgemeine Untersuchung über die Newtonsche Theorie der Fernwirkung,* B. G. Teubner, Leipzig.

NEWCOMB, S. (1895). "The elements of the four inner planets and the fundamental constants of astronomy"—Supplement to the *American Ephemeris and Nautical Almanacs for 1897,* Government Printing Office, Washington.

NEWTON, I. (1687). *Philosophiae Naturalis Principia Mathematica,* London; see also *Isaac Newton's Philosophiae Naturalis Principia Mathematica,* Vols. I and II, eds. A. Koyré and J. B. Cohen, University Press, Cambridge (1972).

OPPENHEIM, S. (1920). *Kritik des Newtonschen Gravitationsgesetzes,* in *Enzyklopädie der Mathematischen Wissenschaften,* Vol. VI, 2nd ed., B. G. Teubner, Leipzig.

RIEMANN, B. (1875). *Schwere, Elektricität und Magnetismus,* C. Rümpler, Hannover.

SCHWARZSCHILD, K. (1916). *Sitzungsber. Preuss. Akad. Wiss.,* 189.

VON SEELIGER, H. (1906). *Münchner Ber.* **36**, 595.

TISSERAND, M. F. (1872). "Sur le mouvement des planètes autour du Soleil d'après la loi électrodynamique de Weber," *C. R. Acad. Sci.* **175**, 760.

References

TISSERAND, M. F. (1895, 1896). *Traité de Mécanique Céleste*, Vols. I–V, Gauthier-Villars, Paris.

TREDER, H.-J. (1972). *Die Relativität der Trägheit*, Akademie-Verlag, Berlin.

TREDER, H.-J. (1975). *Ann. Phys.* (*Leipzig*) **32**, 338.

WIECHERT, E. (1916). *Phys. Zeitschr.* **17**, 442.

WIECHERT, E. (1925). "Die Mechanik im Rahmen der Allgemeinen Physik," in E. Lecher (ed.), *Physik*, B. G. Teubner, Leipzig.

ZÖLLNER, C. F. (1872). *Die Natur der Cometen* (3rd ed. in 1883), L. Staackmann, Leipzig.

2.2. The Significance of Gravitational Optics

Unified Field Theory

EINSTEIN, A. (1950, 1955). *The Meaning of Relativity*, 3rd (1950) and 5th (1955) editions, Princeton University Press. Appendix II (1950): "The Generalized Theory of Gravitation." Appendix II (1955): "The Relativistic Theory of the Non-Symmetric Field."

HLAVATÝ, V. (1957). *Geometry of Einstein's Unified Field Theory*, Noordhoff, Groningen.

SCHRÖDINGER, E. (1950). *Space–Time Structure*, University Press, Cambridge.

TONNELAT, M. A. (1965). *Les Théories Unitaires de l'Électromagnetisme et de la Gravitation*, Gauthiers-Villars, Paris.

Gravito-Optics

FREUNDLICH, E. F. (1929). *Phil. Mag.* **45**, 303.

MERAT, P., PECKER, J. C., VIGIER, J. P., AND YOURGRAU, W. (1974). *Astron. Astrophys.* **32**, 471.

References

MIKHAILOV, A. A. (1959). *Mon. Not. R. Astron. Soc.* **119**, 593.

TREDER, H.-J. (1971). *Ann. Phys.* (*Leipzig*) **27**, 177.

WOODWARD, J. F. AND YOURGRAU, W. (1972). *Nuovo Cimento* **9B**, 440.

References for Chapter 3

AMBARTSUMJAN, V. A. (1975). *Sitzungsber. Akad. Wiss. D.D.R.*, 15N.

DICKE, R. H. (1964). "The many faces of Mach," in Chiu, H.-J. and Hoffmann, W. F. (eds.), *Gravitation and Relativity*, W. A. Benjamin, New York–Amsterdam.

EDDINGTON, A. S. (1936). *Relativity Theory of Protons and Electrons*, University Press, Cambridge.

EDDINGTON, A. S. (1948). *Fundamental Theory*, University Press, Cambridge.

EINSTEIN, A. (1912). *Vierteljahresschr. Gerichtl. Med.* **44**, 37.

EINSTEIN, A. (1913). *Phys. Z.* **14**, 1249.

EINSTEIN, A. (1917). *Sitzungsber. Preuss. Akad. Wiss.*, 124.

EINSTEIN, A. (1921). *The Meaning of Relativity*, University Press, Princeton. Later editions 1945, 1956.

EINSTEIN, A. (1922). *Ann. Phys.* (*Leipzig*) **69**, 486.

EINSTEIN, A. (1969). *Grundzüge der Relativitätstheorie*, 5th ed., F. Vieweg, Berlin–Braunschweig.

FOKKER, A. D. (1965). *Time and Space, Weight and Inertia*, Pergamon Press, Elmsford (New York).

FRIEDLÄNDER, B. AND J. (1896). *Absolute oder Relative Bewegung?*, L. Simion, Berlin.

HECKMANN, O. (1969). *Theorien der Kosmologie*, J. Springer, Berlin. First edition 1942.

References

HERTZ, H. (1894). *Die Prinzipien der Mechanik*, J. A. Barth, Leipzig. Later edition 1910.

JORDAN, P. (1961). *Die Expansion der Erde*, F. Vieweg, Braunschweig.

MACH, E. (1883). *Die Mechanik in ihrer Entwicklung*, F. A. Brockhaus, Leipzig. Later edition 1933.

McCREA, W. H. AND MILNE, E. A. (1934). *Quart. J. Math. Oxford* **5**, 73.

MILNE, E. A. (1933). *Relativity, Gravitation and World Structure*, University Press, Oxford.

MILNE, E. A. (1934). *Quart. J. Math. Oxford* **5**, 64.

MILNE, E. A. (1948). *Kinematic Relativity*, University Press, Oxford. Later edition 1951.

MURADJAN, R. M. (1975). *Astrofizika* **11**, 237.

NEUMANN, C. (1870), *Über die Prinzipien der Galilei–Newtonschen Theorie*, B. G. Teubner, Leipzig.

NEUMANN, C. (1896). *Allgemeine Untersuchungen über die Newtonsche Theorie der Fernwirkung*, B. G. Teubner, Leipzig.

PLANCK, M. (1887). *Das Prinzip der Erhaltung der Energie*, B. G. Teubner, Berlin–Leipzig. Later editions 1909, 1913.

POINCARÉ, H. (1912). *Wissenschaft und Hypothese*, 3rd ed., B. G. Teubner, Leipzig.

POINCARÉ, H. (1914). *Wissenschaft und Methode*, B. G. Teubner, Leipzig–Berlin.

RIEMANN, B. (1880). *Schwere, Elektrizität und Magnetismus*, 2nd ed., C. Rümpler, Hannover.

SELETY, F. (1922). *Ann. Phys. (Leipzig)* **68**, 281.

DE SITTER, W. (1917). *Proc. Acad. Sci. Amsterdam* **19**, 1217.

THIRRING, H. (1918). *Phys. Z.* **19**, 33.

THIRRING, H. (1921). *Phys. Z.* **22**, 29.

References

TREDER, H.-J. (1972). *Die Relativität der Trägheit*, Akademie-Verlag, Berlin.

TREDER, H.-J. (1973a). *Gerlands Beitr. Geophys.* **82**, 92.

TREDER, H.-J. (1973b). *Symposia Mathematica* VII, Academic Press, New York–London.

TREDER, H.-J. (1974a). *Astron. Nachr.* **295**, 1 and 55.

TREDER, H.-J. (1974b). *Prinzipien der Dynamik bei Einstein, Hertz, Mach und Poincaré*, Akademie-Verlag, Berlin.

TREDER, H.-J. (1975). *Astron. Nachr.* **296**, 101.

TREDER, H.-J. (1976a). *Astrofizika* **12**, 511.

TREDER, H.-J. (1976b). *Astron. Nachr.* **297**, 113.

WEYL, H. (1923). *Raum–Zeit–Materie*, 5th ed., J. Springer, Berlin.

WEYL, H. (1924). *Naturwissenschaften* **12**, 197.

YOURGRAU, W., AND MANDELSTAM, S. (1960). *Variational Principles in Dynamics and Quantum Theory*, W. B. Saunders, London.

References for Appendix A

For more details on the non-Riemannian representation of space–time geometry, see:

HLAVATÝ, V. (1957). *Geometry of Einstein's Unified Field Theory*, Noordhoff, Groningen.

LEVI-CIVITA, T. (1950). *The Absolute Differential Calculus*, Blackie and Son, London–Glasgow.

MERCIER, A., TREDER, H.-J., AND YOURGRAU, W. (1979). *On General Relativity*, Akademie-Verlag, Berlin.

References for Appendix B

EINSTEIN, A. (1928). *Sitzungsber. Preuss. Akad. Wiss.*, 215.

EINSTEIN, A., AND MAYER, W. (1932). *Sitzungsber. Preuss. Akad. Wiss.*, 522.

EISENHART, L. P. (1927). *Non-Riemannian Geometry*, American Mathematical Society, New York.

INFELD, L., AND VAN DER WAERDEN, B. L. (1933). *Sitzungsber. Preuss. Akad. Wiss.*, 311.

SCHOUTEN, J. A. (1953). *Ricci-Calculus*, J. Springer, Berlin–Göttingen–Heidelberg.

TREDER, H.-J. (1966). "Einstein-Gruppe und Raumstruktur," in *Entstehung, Entwicklung und Perspektiven der Einsteinschen Gravitationstheorie*, Akademie-Verlag, Berlin.

TREDER, H.-J. (1967). *Ann. Phys. (Leipzig)* **20**, 194.

TREDER, H.-J. (1969). *Math. Nachr.* **33**, 295.

TREDER, H.-J. (1971). *Monatsber. Deutsch. Akad. Wiss.* **13**, 310.

WEYL, H. (1929). *Z. Phys.* **56**, 330.

Author Index

209

Author Index

Subject Index

Subject Index

Subject Index

Subject Index

Subject Index

Subject Index